T0335694

Earthquakes, Tsunamis and Nuclear Risks

Katsuhiro Kamae

Editor

Earthquakes, Tsunamis and Nuclear Risks

Prediction and Assessment Beyond the
Fukushima Accident

 Springer Open

Editor
Katsuhiro Kamae
Research Reactor Institute
Kyoto University
Kumatori, Osaka, Japan

ISBN 978-4-431-55820-0   ISBN 978-4-431-55822-4   (eBook)
DOI 10.1007/978-4-431-55822-4

Library of Congress Control Number: 2015957422

Springer Tokyo Heidelberg New York Dordrecht London
© The Author(s) 2016. The book is published with open access at SpringerLink.com.

Printed on acid-free paper

Springer Japan KK is part of Springer Science+Business Media (www.springer.com)

Foreword

On March 11, 2011, a massive earthquake and the resultant tsunami struck the Tohoku area of Japan, causing serious damage to the Fukushima Daiichi Nuclear Power Plant and the release of a significant quantity of radionuclides into the surrounding environment. This accident underlined the necessity of establishing more comprehensive scientific research for promoting safety in nuclear technology. In this situation, the Kyoto University Research Reactor Institute (KURRI) established a new research program called the "KUR Research Program for Scientific Basis of Nuclear Safety" in 2012.

Nuclear safety study includes not only the prevention of nuclear accidents but also the safety measures after the accident from a wider viewpoint ensuring the safety of residents. A long time is needed for the improvement of the situation, but the social needs for the reinforcement of nuclear safety increases rapidly. The advancement of disaster prevention technology for natural disasters such as earthquakes and tsunamis, the reinforcement of measures for the influence of accidents, and the reinforcement of the safety management of spent fuels and radioactive wastes are demanded, not to mention the reinforcement of nuclear reactor safety. Also demanded are the underlying mechanism investigation and accurate assessment for the effect of radiation on the human body and life. As with all premises, detailed inspection and analysis of the accident are indispensable.

In the Research Program for the Scientific Basis of Nuclear Safety, an annual series of international symposia was planned along with specific research activities. The first in the series of symposia, entitled "The International Symposium on Environmental Monitoring and Dose Estimation of Residents after the Accident of TEPCO's Fukushima Daiichi Nuclear Power Stations [sic]," was held on December 14, 2012, concerning the radiological effects of the accident on the public. Also the second was held on November 28, 2013, dealing with nuclear back-end issues and the role of nuclear transmutation technology after the accident of TEPCO's Fukushima Daiichi NPP. The results covering a wide range of research activities were reported and discussed in the symposium and were published at the request of many people, with open access.

Following these symposia, the third annual symposium in this series, entitled "Earthquake, Tsunami and Nuclear Risks after the Accident of TEPCO's Fukushima Daiichi Nuclear Power Stations," was held on October 30, 2014, related to the safety evaluation of nuclear power plants which was being performed in Japan for natural external hazards. The regulatory framework has been based on the deterministic approach so far, but it has been pointed out that many uncertainties in natural external hazards such as seismic motion and tsunami should be considered. Thus the symposium has dealt with the uncertainties in the safety evaluation, probabilistic risk assessment of earthquakes and tsunamis, fault displacement hazard evaluation, risk of nuclear systems, risk communication, and so on.

This publication summarizes the current status of the methodology for the assessment of nuclear risks of serious consequence, low-probability events, which has been presented and discussed during the symposium. It will contribute to better understanding and further discussion of the issues.

On behalf of KURRI, I wish to thank all the contributors to this publication as well as the reviewers. KURRI also hopes that this publication will promote further progress in nuclear safety research and will contribute to the reduction of public anxiety after the accident.

Hirotake Moriyama
Kyoto University Research Reactor Institute
Osaka, Japan

Preface

The accident at the Fukushima Daiichi Nuclear Power Plant, which occurred in March 2011, has strengthened the safety evaluation of nuclear power plants in cases of natural disasters. New safety regulations have been instituted on the basis of a deterministic approach aimed at absolute safety, such as the requirement of safe sites in the case of tsunamis and the prohibition of installation of especially important facilities on the outcrop of an active fault. Because many uncertainties exist in natural phenomena such as earthquakes and tsunamis, a risk concept for seismic motion and tsunamis beyond a design basis is indispensable.

Under such circumstances, an international symposium "Earth quake, Tsunami and Nuclear Risks After the accident of TEPCO's Fukushima Daiichi Nuclear Power Stations" was held in Kyoto, Japan, on October 30, 2014. This symposium was hosted by the Kyoto University Research Reactor Institute under the "KUR Research Program for the Scientific Basis of Nuclear Safety."

The topics of the symposium included uncertainties in the safety evaluation of an earthquake and tsunami, probabilistic risk assessment (PRA) for earthquake and tsunami, risk in a nuclear system, and risk communication.

This book includes some of the presentations at the symposium. The main topics of the book are (1) Active faults and active tectonics important for seismic hazard assessment of nuclear facilities, (2) Seismic source modeling, simulation, and modeling techniques indispensable for strong ground motion prediction, and (3) PRA with external hazards and risk communication.

All the articles in this book were peer-reviewed by specialists in the relevant fields and are listed in the Contents as Cooperators. The editor would like to thank all the authors and these cooperating specialists who worked so hard to publish this book. I hope this volume will provide readers the opportunity to consider the future direction of nuclear safety in the face of natural disasters.

Osaka, Japan   Katsuhiro Kamae

Cooperators

Tomotaka Iwata (DPRI, Kyoto University)
Hirotoshi Uebayashi (Research Reactor Institute, Kyoto University)
Michihiro Ohori (University of Fukui)
Takashi Kumamoto (Okayama University)
Hideaki Goto (Hirosima University, Hiroshima Japan)
Tsuyoshi Takada (University of Tokyo, Tokyo Japan)
Prof. Osamu Furuya (Tokyo City University, Tokyo Japan)
Nobuhisa Matsuta (Okayama University, Okayama Japan)
Hiroshi Miyano (Hosei University)
Ken Muramatsu (Tokyo City University, Tokyo Japan)
Akira Yamaguchi (The University of Tokyo, Ibaraki Japan)

Contents

Part I Active Faults

1   Examination of the Correlation Between Tectonic Landforms and
    Shallow Subsurface Structural Datasets for the Estimation of
    Seismic Source Faults.................................   3
    Takashi Kumamoto, Masatoshi Fujita, Hideaki Goto,
    and Takashi Nakata

2   Multivariate Statistical Analysis for Seismotectonic Provinces
    Using Earthquake, Active Fault, and Crustal Structure Datasets...   31
    Takashi Kumamoto, Masataka Tsukada, and Masatoshi Fujita

3   Multiple Regression Analysis for Estimating Earthquake
    Magnitude as a Function of Fault Length and Recurrence
    Interval..   43
    Takashi Kumamoto, Kozo Oonishi, Yoko Futagami,
    and Mark W. Stirling

4   Coseismic Tsunami Simulation Assuming the Displacement
    of High-Angle Branching Active Faults Identified on the
    Continental Slope Around the Japan Trench................   55
    Shota Muroi and Takashi Kumamoto

5   Extensive Area of Topographic Anaglyphs Covering Inland and
    Seafloor Derived Using a Detailed Digital Elevation Model for
    Identifying Broad Tectonic Deformations..................   65
    Hideaki Goto

Part II   Seismic Source Modeling and Seismic Motion

6   **Relation Between Stress Drops and Depths of Strong Motion
    Generation Areas Based on Previous Broadband Source Models
    for Crustal Earthquakes in Japan**   77
    Toshimi Satoh and Atsushi Okazaki

7   **Heterogeneous Dynamic Stress Drops on Asperities in Inland
    Earthquakes Caused by Very Long Faults and Their Application
    to the Strong Ground Motion Prediction**   87
    Kazuo Dan, Masanobu Tohdo, Atsuko Oana, Toru Ishii,
    Hiroyuki Fujiwara, and Nobuyuki Morikawa

8   **Simulation of Broadband Strong Motion Based on the Empirical
    Green's Spatial Derivative Method**   99
    Michihiro Ohori

Part III   Probabilistic Risk Assessment with External Hazards

9   **Development of Risk Assessment Methodology Against External
    Hazards for Sodium-Cooled Fast Reactors** 111
    Hidemasa Yamano, Hiroyuki Nishino, Yasushi Okano,
    Takahiro Yamamoto, and Takashi Takata

10 **Effectiveness Evaluation About the Tsunami Measures Taken at
    Kashiwazaki-Kariwa NPS** 123
    Masato Mizokami, Takashi Uemura, Yoshihiro Oyama,
    Yasunori Yamanaka, and Shinichi Kawamura

11 **Development of a New Mathematical Framework for Seismic
    Probabilistic Risk Assessment for Nuclear Power Plants – Plan
    and Current Status –** 137
    Hitoshi Muta, Ken Muramatsu, Osamu Furuya, Tomoaki Uchiyama,
    Akemi Nishida, and Tsuyoshi Takada

Part IV   Nuclear Risk Governance in Society

12 **Deficits of Japanese Nuclear Risk Governance Remaining After
    the Fukushima Accident: Case of Contaminated Water
    Management** 157
    Kohta Juraku

13 **A Community-Based Risk Communication Approach on Low-Dose
    Radiation Effect** 171
    Naoki Yamano

Part I
Active Faults

Chapter 1
Examination of the Correlation Between Tectonic Landforms and Shallow Subsurface Structural Datasets for the Estimation of Seismic Source Faults

Takashi Kumamoto, Masatoshi Fujita, Hideaki Goto, and Takashi Nakata

Abstract Estimation of the magnitudes of future earthquakes produced by faults is critical in seismic hazard assessment, especially for faults that are short in extent compared with the thickness of the seismogenic layers of the upper crust. A new seismogenic fault model for earthquake size estimation was constructed by combining (a) new assessments of the precise location and distribution of active faults from aerial photograph analysis and (b) estimations of subsurface structures from geological, gravity, and seismicity datasets. The integrated results of (1) tectonic landforms determined from aerial photographs, (2) geologic data showing the distribution of geologic faults, (3) Bouguer gravity anomaly data over wavelengths of 4–200 km, and (4) seismicity data were superimposed on geographic information system (GIS) data around the nuclear power plants in Japan. The results indicate the possible occurrence of large earthquakes, because the lengths of the subsurface earthquake faults were estimated to be longer than the length of the surface faults if subsurface structures were included.

Keywords Aerial photograph analysis • Distribution of active faults • Subsurface structure

T. Kumamoto (✉)
Okayama University, Fact. Science, Tsushimanaka 3-1-1, Kita-ku, Okayama 700-8530, Japan
e-mail: tkuma@cc.okayama-u.ac.jp

M. Fujita
Nuclear Regulation Authority, Roppongi 1-9-9, Minato-ku, Tokyo 106-8450, Japan

H. Goto • T. Nakata
Hiroshima University, Fact. Letter, Kagamiyama 1-3-2, Higashi-Hiroshima 739-8511, Japan

K. Kamae (ed.), *Earthquakes, Tsunamis and Nuclear Risks*,
DOI 10.1007/978-4-431-55822-4_1

3

1.1   Introduction

The Headquarters for Earthquake Research Promotion (HERP) published the "Method of long-term evaluation of active fault (preliminary version)" (HERP [1]), a new integrated method of active fault assessment for seismic hazard analysis. Several problems were highlighted in this report, with two being of particular importance. First, the surface ruptures of some recent intraplate earthquakes in Japan were shorter than the source fault of the earthquake in the subsurface. Second, it is necessary to update Matsuda's "5 km rule" [2], the current reference criterion of fault gap distance for assessing if neighboring faults rupture simultaneously, which is widely used for seismic hazard analysis in Japan.

The HERP [1] also included some ideas for improvement and methods to solve the problem of the mismatch between surface and subsurface fault lengths for moderate to large earthquakes. One example is to incorporate subsurface structural datasets such as geologic maps, gravity anomaly data, and instrumentally observed seismicity data with the surface distribution of active faults deduced from aerial photograph analysis to better estimate subsurface earthquake source faults.

Preliminary results of the new method of comparing surface and subsurface structures are outlined here, especially for the areas near nuclear power plants in Japan. We conducted aerial photograph analysis to identify tectonic landforms and created fault distribution maps. Datasets representing subsurface structures in the study areas were overlain on the maps using geographic information system (GIS) techniques. We then estimated (a) the length of earthquake source fault for isolated faults with short surface lengths and (b) the possibility that neighboring surface faults would rupture simultaneously due to subsurface continuity.

1.2   Data

The reference criteria and legends for the aerial photograph analysis in this study are the same as those in the "Active Fault Map in Urban Area" published by the Geospatial Information Authority of Japan [3], and cross-check rule is also applied for the analysis. The top section of Fig. 1.1 (1)–(14) shows the distribution of active faults in the study areas determined by aerial photograph analysis. The rectangles in each section indicate a second-order map grid (scale 1:25,000) from the Geospatial Information Authority of Japan and show the areas of detailed aerial photograph analysis in this study. The numbers of the top sections in Fig. 1.1 indicate the corresponding identification in Table 1.1. The middle left section shows the distribution of "Active faults in Japan" (RGAF [4]), and the middle right section shows the distribution in the "Digital Active Fault Map of Japan" (Nakata and Imaizumi ed. [5]) for comparison. As the same criterion was used for the "Digital Active Fault Map of Japan" and this study, there is little difference in the results.

A total of 249 active faults were identified in this study (Table 1.1), with 78 of these partially or completely corresponding to faults identified in "Active faults in Japan" (RGAF [4]). Among the remaining 171 faults, 164 were 10 km or shorter in

(1) Tomari

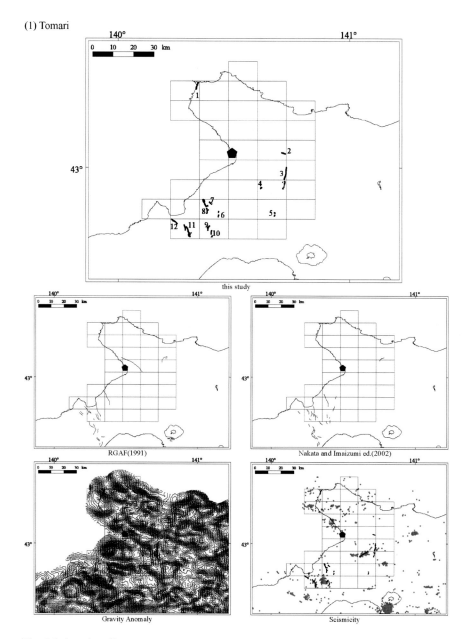

Fig. 1.1 (continued)

(2) Oma

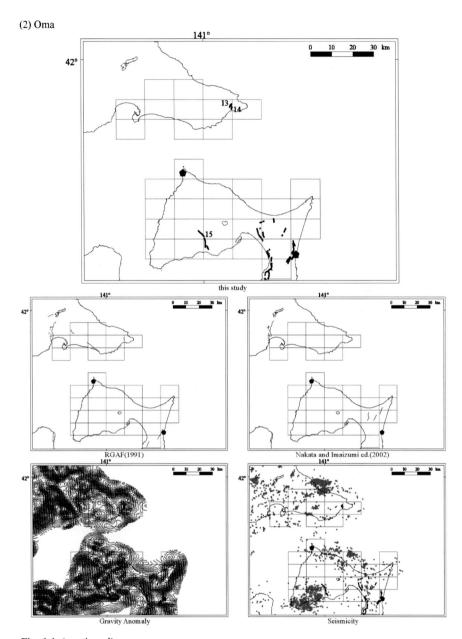

Fig. 1.1 (continued)

(3) Higashidori

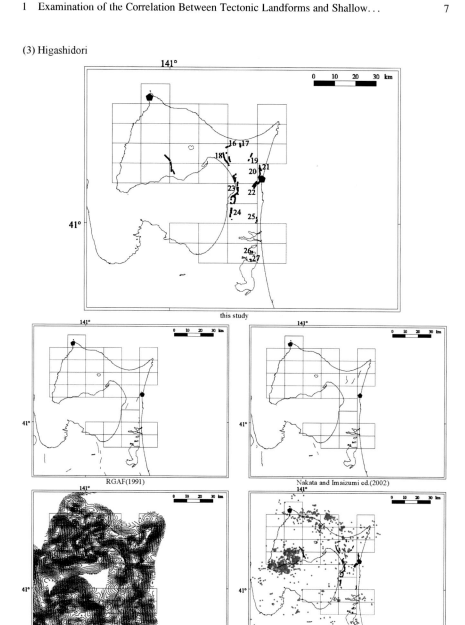

Fig. 1.1 (continued)

(4) Onagawa

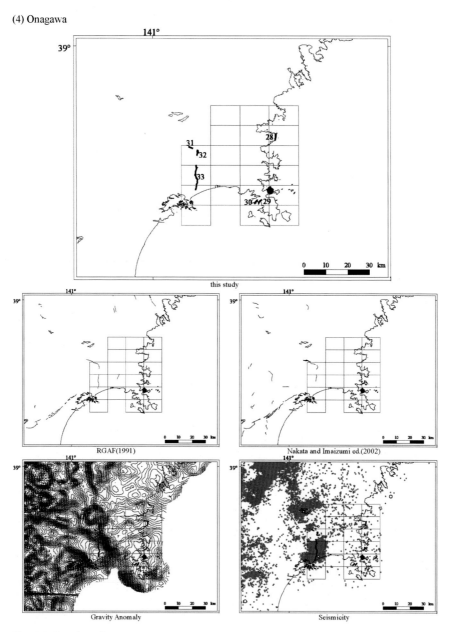

Fig. 1.1 (continued)

(5) Fukushima

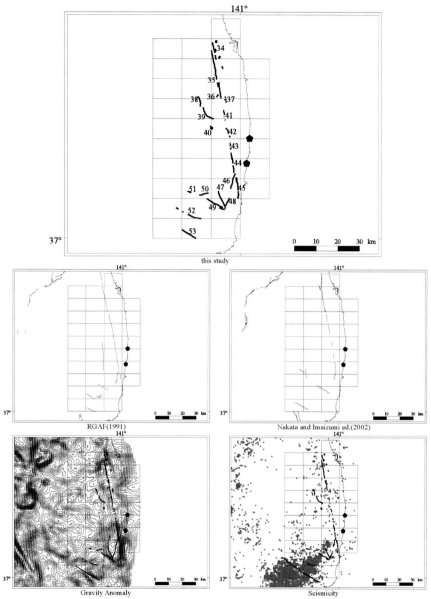

Fig. 1.1 (continued)

(6) Tokai

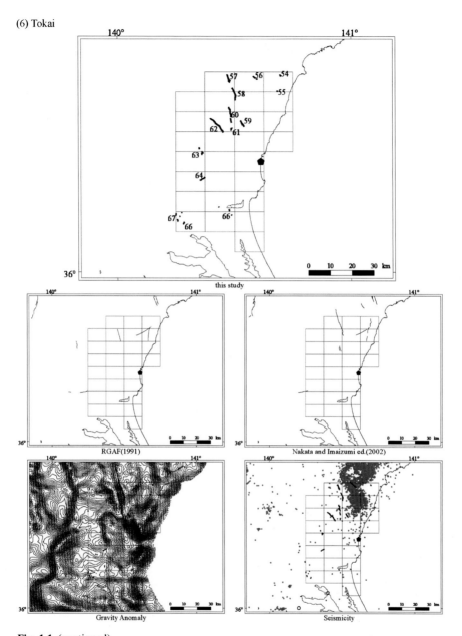

Fig. 1.1 (continued)

(7) Kashiwazaki

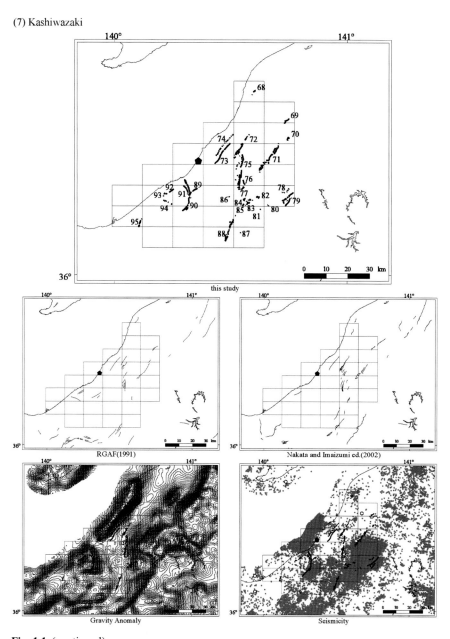

Fig. 1.1 (continued)

(8) Hamaoka

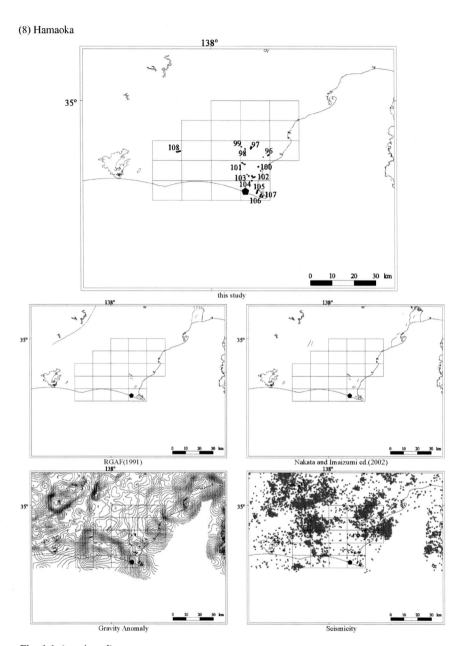

Fig. 1.1 (continued)

(9) Shika

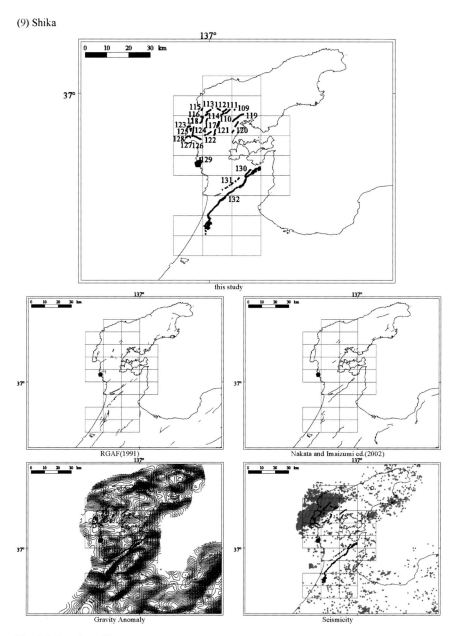

Fig. 1.1 (continued)

14

T. Kumamoto et al.

(10) Tsuruga

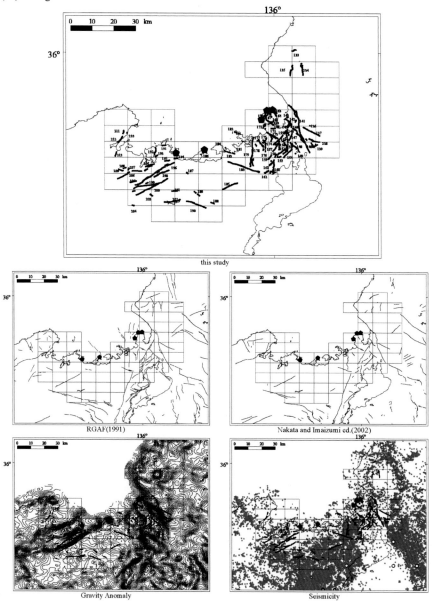

Fig. 1.1 (continued)

(11) Shimane

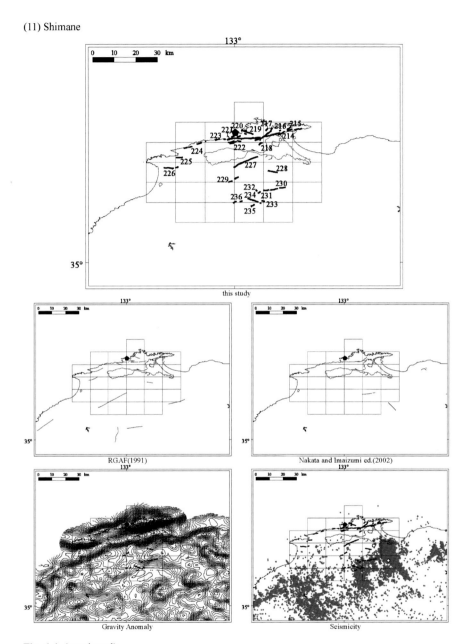

Fig. 1.1 (continued)

(12) Ikata

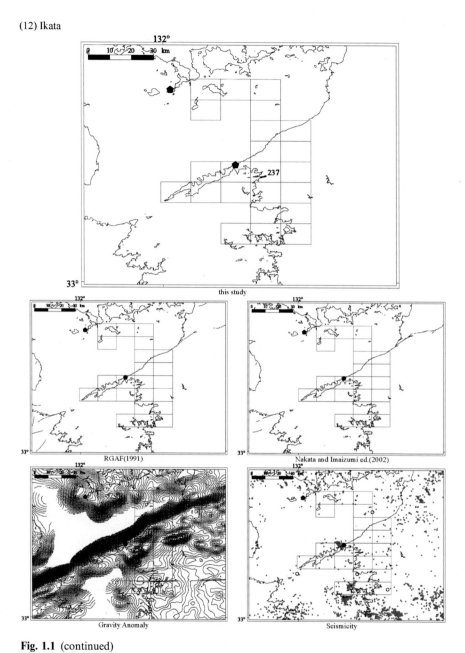

Fig. 1.1 (continued)

(13) Genkai

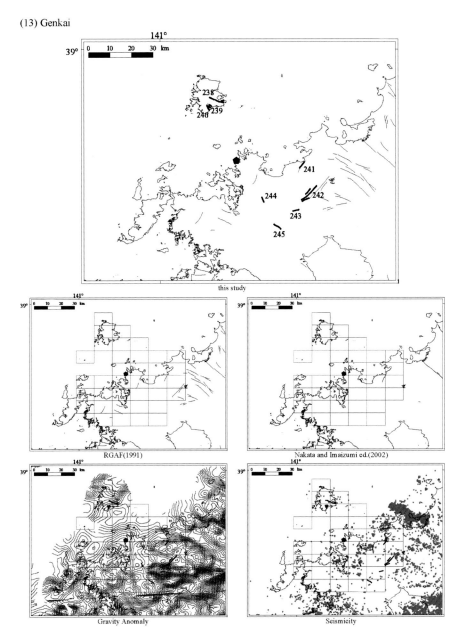

Fig. 1.1 (continued)

(14) Sendai

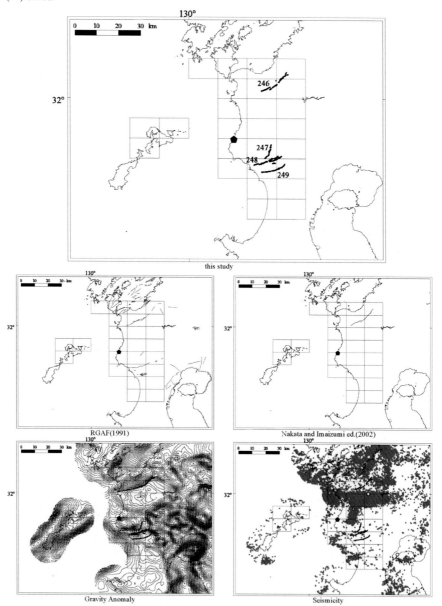

Fig. 1.1 Distribution of active faults (*top*: this study, *middle left*: RGAF (1991), *middle right* (Nakata and Imaizumi [5])) and subsurface datasets (*bottom left*: 1 mgal contour of Bouguer gravity anomaly, *bottom right*: observed seismicity) of 14 subjected areas

Table 1.1 Surface and subsurface fault length in Fig. 1.1 and Fig. 1.2

Site	ID	This study	RGAF (1991)	Geology	Gravity	Seismicity
Tomari	1	3.6	–	5.0	–	–
Tomari	2	1.7	–	3.4	–	–
Tomari	3	8.8	–	14.4	14.4	–
Tomari	4	0.9	–	–	–	6.3
Tomari	5	1.3	–	–	–	–
Tomari	6	1.6	–	3.5	–	–
Tomari	7	1.3	–	–	–	–
Tomari	8	6.4	12.0	–	15.5	–
Tomari	9	2.6	–	5.7	–	–
Tomari	10	2.5	–	–	–	–
Tomari	11	4.7	6.0	–	–	–
Tomari	12	3.2	6.0	4.7	7.1	–
Oma	13	1.6	3.0	–	–	–
Oma	14	1.2	3.0	3.5	3.5	–
Oma	15	7.6	–	10.8	–	–
Higashidori	16	1.7	–	–	–	–
Higashidori	17	2.0	–	–	–	–
Higashidori	18	5.6	–	25.6	26.7	–
Higashidori	19	2.7	–	–	–	–
Higashidori	20	6.1	–	–	–	–
Higashidori	21	1.5	–	–	–	–
Higashidori	22	3.9	7.0	–	–	–
Higashidori	23	9.9	4.0	25.6	26.7	–
Higashidori	24	3.9	–	–	–	–
Higashidori	25	1.5	4.0	4.1	–	–
Higashidori	26	1.0	–	–	–	–
Higashidori	27	1.2	–	–	–	–
Onagawa	28	3.2	–	–	–	–
Onagawa	29	2.1	–	4.4	–	–
Onagawa	30	2.0	–	–	–	–
Onagawa	31	1.8	12.0	5.6	–	–
Onagawa	32	2.8	12.0	8.5	8.5	–
Onagawa	33	12.3	8.0	12.5	–	16.5
Fukushima	34	7.3	–	12.7	–	–
Fukushima	35	20.3	55.0	67.5	36.1	–
Fukushima	36	1.5	–	–	–	–
Fukushima	37	1.7	–	–	–	–
Fukushima	38	4.9	–	–	–	–
Fukushima	39	7.1	–	–	–	–
Fukushima	40	1.7	–	–	–	–
Fukushima	41	2.1	55.0	67.5	36.1	–
Fukushima	42	3.3	55.0	67.5	–	–

(continued)

Table 1.1 (continued)

Site	ID	This study	RGAF (1991)	Geology	Gravity	Seismicity
Fukushima	43	1.6	55.0	67.5	–	–
Fukushima	44	8.1	55.0	67.5	25.8	–
Fukushima	45	8.3	15.0	13.5	16.9	–
Fukushima	46	9.1	10.0	67.5	25.8	–
Fukushima	47	3.6	–	–	25.8	–
Fukushima	48	8.7	–	–	17.9	–
Fukushima	49	7.6	6.0	8.0	–	–
Fukushima	50	3.6	–	–	–	–
Fukushima	51	1.3	–	–	–	–
Fukushima	52	7.0	5.0	19.3	–	–
Fukushima	53	6.8	6.0	16.9	18.2	28.9
Tokai	54	0.9	–	–	–	–
Tokai	55	0.5	10.0	–	–	–
Tokai	56	1.8	–	–	–	5.6
Tokai	57	3.0	–	–	26.0	–
Tokai	58	6.0	–	–	26.0	–
Tokai	59	3.2	–	–	–	–
Tokai	60	6.8	7.0	11.7	23.9	–
Tokai	61	1.1	–	–	–	–
Tokai	62	8.4	–	12.1	–	–
Tokai	63	1.7	–	–	–	–
Tokai	64	2.2	–	10.3	–	–
Tokai	65	0.6	–	–	–	–
Tokai	66	1.1	–	–	–	–
Tokai	67	2.0	–	–	–	–
Kashiwazaki	68	0.8	–	–	12.6	–
Kashiwazaki	69	2.1	4.0	3.2	–	–
Kashiwazaki	70	1.5	–	–	–	–
Kashiwazaki	71	11.3	15.0	16.8	–	–
Kashiwazaki	72	6.4	5.0	14.5	–	–
Kashiwazaki	73	10.0	–	14.4	–	–
Kashiwazaki	74	5.0	11.0	13.3	31.8	–
Kashiwazaki	75	11.5	7.0	16.2	16.2	–
Kashiwazaki	76	1.8	–	–	–	–
Kashiwazaki	77	1.5	–	–	–	–
Kashiwazaki	78	1.0	–	–	–	–
Kashiwazaki	79	4.6	–	9.0	–	9.6
Kashiwazaki	80	0.3	–	–	–	–
Kashiwazaki	81	0.3	–	–	–	–
Kashiwazaki	82	0.7	–	–	–	–
Kashiwazaki	83	1.1	3.0	–	8.9	–
Kashiwazaki	84	1.0	–	–	–	–

(continued)

Table 1.1 (continued)

Site	ID	This study	RGAF (1991)	Geology	Gravity	Seismicity
Kashiwazaki	85	1.0	–	–	–	–
Kashiwazaki	86	0.5	–	1.8	–	–
Kashiwazaki	87	0.2	–	–	–	–
Kashiwazaki	88	7.6	–	12.2	12.2	–
Kashiwazaki	89	2.2	–	–	–	–
Kashiwazaki	90	3.3	–	–	–	–
Kashiwazaki	91	13.3	–	–	–	20.2
Kashiwazaki	92	2.7	5.0	5.8	5.4	–
Kashiwazaki	93	0.8	–	–	–	–
Kashiwazaki	94	1.3	–	4.4	–	–
Kashiwazaki	95	2.9	–	6.7	–	–
Hamaoka	96	1.2	–	–	–	–
Hamaoka	97	1.6	–	–	–	–
Hamaoka	98	0.6	–	–	–	–
Hamaoka	99	0.3	–	–	–	–
Hamaoka	100	0.7	–	–	–	–
Hamaoka	101	1.0	2.0	–	–	–
Hamaoka	102	1.3	3.0	–	–	–
Hamaoka	103	0.7	–	–	–	–
Hamaoka	104	0.5	–	–	–	–
Hamaoka	105	1.9	3.0	–	–	–
Hamaoka	106	1.5	2.0	–	–	–
Hamaoka	107	0.8	1.0	–	–	–
Hamaoka	108	1.9	–	–	–	–
Shika	109	0.5	–	–	–	–
Shika	110	4.3	2.0	7.4	–	–
Shika	111	2.3	–	–	–	–
Shika	112	1.6	–	–	–	–
Shika	113	2.6	–	–	–	6.4
Shika	114	1.5	–	–	–	–
Shika	115	1.3	–	–	–	–
Shika	116	1.1	–	–	–	2.8
Shika	117	1.5	–	–	–	–
Shika	118	3.4	–	–	–	4.6
Shika	119	5.7	–	–	–	–
Shika	120	3.9	3.0	6.2	–	–
Shika	121	6.3	2.0	–	–	–
Shika	122	2.8	–	–	–	–
Shika	123	1.3	–	–	–	–
Shika	124	4.7	4.0	7.4	–	–
Shika	125	1.5	–	–	–	–
Shika	126	3.6	–	–	–	–

(continued)

Table 1.1 (continued)

Site	ID	This study	RGAF (1991)	Geology	Gravity	Seismicity
Shika	127	1.0	2.0	–	–	–
Shika	128	1.1	–	–	–	–
Shika	129	1.9	3.0	–	–	–
Shika	130	3.8	3.0	24.6	24.6	–
Shika	131	3.7	9.0	24.6	24.6	–
Shika	132	33.3	45.0	50.2	50.2	–
Tsuruga	133	2.5	12.0	11.6	–	–
Tsuruga	134	7.0	–	–	9.6	–
Tsuruga	135	5.7	10.0	7.2	–	–
Tsuruga	136	0.5	–	–	–	–
Tsuruga	137	8.7	13.0	–	–	–
Tsuruga	138	5.0	–	–	–	–
Tsuruga	139	2.0	–	–	4.4	–
Tsuruga	140	3.2	–	–	11.7	–
Tsuruga	141	23.4	37.0	33.9	35.6	–
Tsuruga	142	1.8	–	–	–	–
Tsuruga	143	1.0	–	–	–	–
Tsuruga	144	12.9	4.0	–	18.1	–
Tsuruga	145	4.4	–	–	–	–
Tsuruga	146	2.3	–	–	–	–
Tsuruga	147	3.5	–	–	–	–
Tsuruga	148	1.2	–	–	–	–
Tsuruga	149	6.5	13.0	9.6	12.2	–
Tsuruga	150	13.3	–	–	–	–
Tsuruga	151	3.8	–	–	–	–
Tsuruga	152	7.0	9.0	–	–	–
Tsuruga	153	4.3	–	–	–	–
Tsuruga	154	6.2	11.0	–	–	–
Tsuruga	155	8.6	–	–	–	–
Tsuruga	156	4.4	6.0	–	–	–
Tsuruga	157	4.5	25.0	8.5	–	–
Tsuruga	158	2.1	–	–	–	–
Tsuruga	159	2.8	25.0	5.1	–	–
Tsuruga	160	6.9	7.0	–	18.5	–
Tsuruga	161	3.6	7.0	–	–	–
Tsuruga	162	2.5	–	8.0	–	–
Tsuruga	163	2.4	3.0	–	–	–
Tsuruga	164	2.4	–	–	–	–
Tsuruga	165	1.2	–	–	–	–
Tsuruga	166	0.9	–	–	–	–
Tsuruga	167	4.6	–	–	–	–
Tsuruga	168	6.7	4.0	–	–	–

(continued)

Table 1.1 (continued)

Site	ID	This study	RGAF (1991)	Geology	Gravity	Seismicity
Tsuruga	169	3.0	–	–	–	–
Tsuruga	170	2.3	–	–	–	–
Tsuruga	171	1.8	–	–	–	–
Tsuruga	172	1.6	–	–	–	–
Tsuruga	173	5.3	6.0	–	–	–
Tsuruga	174	2.5	–	–	–	–
Tsuruga	175	1.9	–	2.6	–	–
Tsuruga	176	5.7	10.0	–	–	–
Tsuruga	177	8.0	7.0	–	–	–
Tsuruga	178	1.5	10.0	–	–	–
Tsuruga	179	14.8	15.0	18.9	–	–
Tsuruga	180	17.5	9.0	20.9	–	–
Tsuruga	181	0.4	–	–	–	–
Tsuruga	182	2.2	–	–	–	–
Tsuruga	183	3.1	–	–	–	–
Tsuruga	184	0.3	–	–	–	–
Tsuruga	185	8.0	–	–	–	11.6
Tsuruga	186	2.5	–	–	–	–
Tsuruga	187	1.3	–	–	–	–
Tsuruga	188	2.2	–	–	–	–
Tsuruga	189	3.9	–	–	–	–
Tsuruga	190	9.8	–	–	–	–
Tsuruga	191	2.1	–	–	–	–
Tsuruga	192	4.4	–	4.9	–	–
Tsuruga	193	1.4	–	–	–	–
Tsuruga	194	1.9	–	–	–	–
Tsuruga	195	2.7	–	3.3	–	–
Tsuruga	196	23.5	–	28.1	27.5	–
Tsuruga	197	3.8	–	–	–	–
Tsuruga	198	8.7	–	13.9	16.7	–
Tsuruga	199	14.4	–	–	15.6	–
Tsuruga	200	18.1	15.0	21.0	31.8	–
Tsuruga	201	2.6	–	–	–	–
Tsuruga	202	7.9	–	–	16.0	–
Tsuruga	203	1.8	–	–	–	–
Tsuruga	204	1.4	30.0	14.0	–	–
Tsuruga	205	2.1	–	–	9.3	–
Tsuruga	206	13.7	–	–	17.5	–
Tsuruga	207	1.5	–	–	–	–
Tsuruga	208	1.1	–	–	–	–
Tsuruga	209	2.3	–	–	–	–
Tsuruga	210	0.7	–	–	–	–

(continued)

Table 1.1 (continued)

Site	ID	This study	RGAF (1991)	Geology	Gravity	Seismicity
Tsuruga	211	1.5	2.0	–	–	–
Tsuruga	212	4.4	20.0	–	–	–
Tsuruga	213	3.0	–	–	–	–
Shimane	214	31.7	22.0	36.8	40.0	–
Shimane	215	2.1	2.0	–	–	–
Shimane	216	0.8	–	–	–	–
Shimane	217	1.3	–	–	–	–
Shimane	218	1.8	–	–	–	–
Shimane	219	4.7	–	–	–	–
Shimane	220	1.5	–	–	–	–
Shimane	221	1.4	–	–	–	–
Shimane	222	4.0	–	–	–	–
Shimane	223	2.7	–	–	–	–
Shimane	224	5.2	–	–	–	–
Shimane	225	2.7	–	–	–	–
Shimane	226	5.0	6.0	10.6	15.0	–
Shimane	227	10.8	–	–	–	–
Shimane	228	3.0	3.0	–	–	–
Shimane	229	3.3	–	–	–	–
Shimane	230	7.2	–	–	–	–
Shimane	231	1.4	–	–	–	–
Shimane	232	0.8	–	–	–	–
Shimane	233	1.3	–	–	–	–
Shimane	234	4.9	–	–	–	–
Shimane	235	1.7	–	–	–	–
Shimane	236	2.2	–	–	–	–
Ikata	237	2.4	4.0	6.8	–	–
Genkai	238	6.6	–	12.5	–	–
Genkai	239	1.4	2.0	–	–	–
Genkai	240	3.0	1.0	–	–	–
Genkai	241	3.8	6.0	–	–	–
Genkai	242	9.2	–	–	–	–
Genkai	243	3.1	–	12.5	–	–
Genkai	244	2.8	–	11.3	–	–
Genkai	245	4.1	–	8.1	–	–
Sendai	246	9.9	–	17.8	29.8	–
Sendai	247	5.6	–	–	–	–
Sendai	248	11.6	14.0	14.5	14.5	–
Sendai	249	9.2	–	11.1	13.3	–

ID corresponds to the top figure in Fig. 1.1

length. Attribute data such as fault slip type, certainty level, and tectonic landforms recognized from aerial photograph analysis are provided in GIS; these digital GIS data will be published via the Internet in the near future.

The geological map used in this study was the "Seamless Geological Map of Japan at a scale of 1:200,000 DVD edition" published by the Geological Survey of Japan [6]. Figure 1.2 shows the composite maps of the study areas superimposed on the fault lines of the top section of Fig. 1.1. The color legend of the geological map is the same as that of the "Seamless Geological Map," but is omitted here due lack of adequate space to include the 387 classifications. Please refer to the original legend. We focused on the correspondence between active surface faults and geological boundaries in Fig. 1.2.

The Bouguer gravity anomaly datasets were taken from the "Gravity CD-ROM of Japan, Ver.2" published by the Geological Survey of Japan [7]. We selected an assumed density of 2.67 g/cm^3 and applied 4–200 km band-pass filter processing to eliminate the effects of long wavelengths due to plate subduction. The contour interval in the bottom left section of Fig. 1.1 is 1 mgal.

The seismicity data for the bottom right sections of Fig. 1.1 were taken from the Japan Meteorological Agency's integrated hypocenter database. Data used were from between 1987 and 2011, to maximize accuracy of depth. We selected earthquakes with depths ≤20 km, within the seismogenic layer of the upper crust, as our focus was on intraplate earthquakes.

The distribution of active faults (top section of Fig. 1.1) was superimposed on the geological map (Fig. 1.2), the gravity anomaly contours (bottom left section of Fig. 1.1), and the shallow seismicity map (bottom right section of Fig. 1.1) to estimate the length of subsurface earthquake faults according to our selected criteria.

1.3 Analysis Methods

The superimposed GIS datasets were visually compared to estimate the length of the subsurface earthquake faults, particularly the extension of short faults on the surface and connections between neighboring faults. Subjectivity is inevitable in visual observation, but in order to be as objective as possible, we applied the following general conceptual criteria.

Figure 1.3 (a) is a schematic diagram detailing how to judge the correspondence between active faults identified by aerial photograph analysis and geological boundaries in the geological map. Attention was paid to the accuracy of these maps because of the map scale difference. If the surface fault line (solid line) matches the location and strike of the geological boundary and the latter is longer, the length of the subsurface earthquake fault is estimated from the length of a series of geological boundaries (dashed line).

Figure 1.3 (b) shows a schematic model for assessing the correspondence between active faults identified by aerial photograph analysis and gravity anomaly data. If the surface fault line is situated in an area of dense contour line distribution (where a change in subsurface structure is interpreted), the length of the subsurface

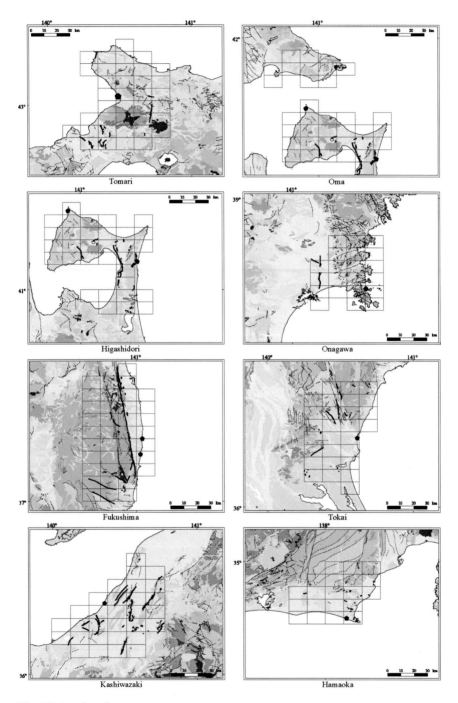

Fig. 1.2 (continued)

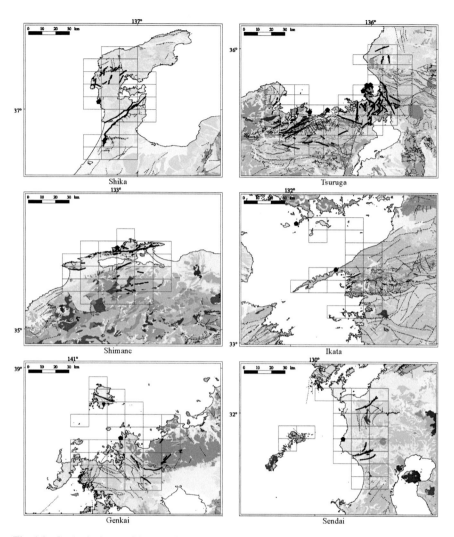

Fig. 1.2 Geological map (GSJ [6]) of 14 subjected areas overlaid with distribution of active faults of this study

earthquake fault (solid line) is estimated from the length of a portion of the dense contour line distribution (dashed line).

Figure 1.3 (c) shows a schematic diagram for determining the correspondence between active faults identified by aerial photograph analysis and seismicity data. If the surface fault line (solid line) is situated in an area with a dense distribution of seismicity, the length of the subsurface earthquake fault is estimated from the length of a portion of a series of seismicity data (dashed line).

Fig. 1.3 Schematic
diagram how to judge the
correspondence between
surface active fault (*solid
line*) and estimated
subsurface earthquake fault
(*dashed line*). *Top*,
geological map; *middle*,
gravity anomaly contour;
bottom, seismicity

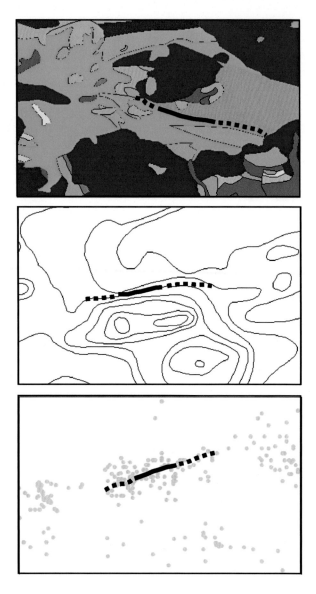

1.4 Results and Discussion

Total 230 active faults with 10 km or shorter length in Table 1.1 were subjects in
discussion since these faults were excluded from the National Seismic Hazard Maps
for Japan by HERP. Among these, 79 faults showed an increase in subsurface
length due to linkages with neighboring faults, and the remaining 151 showed no
clear correspondence. Table 1.1 shows the identification in Fig. 1.1, the lengths of

the surface faults in this study, the lengths in of the faults in the "Active Fault Book," and the estimated lengths in this study from comparisons with the geological map, gravity anomaly contour data, and seismicity data.

The results of this study indicate that the average ratio and standard deviation between the lengths of active faults estimated on the surface and the subsurface is 7.1 and 7.3 for the geologic data, 8.4 and 4.2 for the gravity data, and 4.7 and 3.3 for the seismicity data, respectively. Although the variances are large, still much attention should be paid to this surface and subsurface structural relation to compensate for small amount of surface displacement due to short active faults. In addition, this new approach might effect the so-called "5 km rule" (Matsuda [2]) for grouping and linking active fault strands on the distribution maps since neighboring faults on surface with 5 km or longer gap/step may become connected if subsurface structures show a series of continuity.

Figure 1.4 shows an example of the relationships between the lengths of the surface fault determined by aerial photograph analysis and the estimated subsurface fault lengths for the three datasets in Table 1.1. This figure shows that (1) the estimated length of the subsurface fault is longer than the length of the surface fault, (2) the variance increases if the surface fault length shortens, (3) there is no clear correlation between surface and subsurface fault length for those faults with 10 km or shorter surface lengths, and (4) the maximum estimated subsurface fault length is approximately 30 km, that is, twice of seismogenic layer (upper crust). This indicates that faults with short surface length compared to the width of seismogenic layer still have the potential to produce large earthquakes. Though the potential earthquake magnitude of active faults with short surface length cannot be estimated solely by the surface fault length, Fig. 1.4 shows some clues of maximum length of

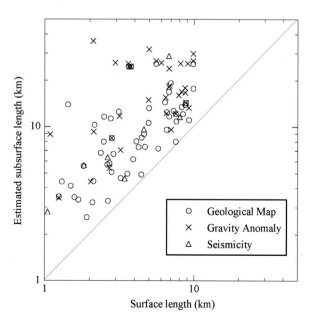

Fig. 1.4 Relation between the length of surface fault from aerial photograph analysis and estimated subsurface fault length among three subsurface datasets in Table 1.1

subsurface fault of 30 km which might relate to both the width of seismogenic layer and the aspect ratio of fault plane.

1.5   Future Challenges

Because aerial photographs were the primary data source for determining surface faults in this study, the study was limited to onshore active faults; no offshore active faults were considered. However, the extension of onshore faults offshore is crucial for seismic hazard assessments for nuclear power plants in Japan. An example is the Tomari nuclear power plant in Hokkaido, where the possibility of extension and linkage of the 32 km long onshore Kuromatsunai fault system to neighboring offshore faults results in a total fault length of 164 km (Asahi newspaper, 2012/3/9).

Nakata and Goto [8] demonstrated a new method for recognizing offshore faults by applying digital bathymetry data to create stereoscopic figures resulting in a seamless connection with onshore faults. It is important to develop surface and subsurface datasets for active faults both onshore and offshore for nuclear power plant seismic hazard assessments and to assess the potential of future earthquakes using multiple integrated data sources.

Acknowledgement   This work is supported by Japan Nuclear Energy Safety Organization (2008–2011) and partly Grant-in-Aid for Scientific Research (21510191).

References

1. HERP (2010) Method of long-term evaluation of active fault (preliminary version) (in Japanese). http://www.jishin.go.jp/main/choukihyoka/katsu_hyokashuho/101125katsu_hyokashuho.pdf
2. Matsuda T (1990) Seismic zoning map of Japanese islands, with maximum magnitudes derived from active fault data. Bull Earthq Res Inst 65:289–319
3. Geospatial Information Authority of Japan (2012) Guide for use of active fault map in urban area (in Japanese). http://www.gsi.go.jp/common/000096698.pdf
4. Research Group for Active Faults of Japan (1991) Active faults in Japan – sheet maps and inventories, Revised edn (in Japanese). University of Tokyo Press, Tokyo
5. Nakata T, Imaizumi T (eds) (2002) Digital active fault map of Japan (in Japanese). University of Tokyo Press, Tokyo
6. Geological Survey of Japan (2009) Seamless geological map of Japan at a scale of 1:200,000 DVD edition (in Japanese), GSJ, Tsukuba
7. Geological Survey of Japan (2004) Gravity CD-ROM of Japan, Ver.2 (in Japanese), GSJ, Tsukuba
8. Nakata T, Goto H (2010) Active faults along the Nankai trough deduced from detailed digital bathymetry data (in Japanese). Kagaku 80:852–857

Chapter 2
Multivariate Statistical Analysis for Seismotectonic Provinces Using Earthquake, Active Fault, and Crustal Structure Datasets

Takashi Kumamoto, Masataka Tsukada, and Masatoshi Fujita

Abstract Seismotectonic zonation for seismic hazard assessment of background faults and earthquakes by the Headquarters for Earthquake Research Promotion (HERP [1]) is based on the results of the seismotectonic boundaries of Kakimi et al. [2]. However, several unsolved problems, such as map scale, remain in this approach for better prediction of the magnitude and frequency of blind earthquakes. The aim of this study was to construct a new quantitative and objective seismo-tectonic province map for the main islands of Japan (Honshu) for rational earthquake size estimation of blind faults and earthquakes. The resolution of the map was set as the second-order map grid of ca. 10 by 10 km of the Geographic Survey of Japan. Then, the parameters of (1) observed seismicity, (2) distribution of active faults converted to earthquake moment release rate, (3) width of the seismogenic layer, and (4) Bouguer gravity anomaly were assigned independently to each grid for principal component analysis. The first principal component of the principal analysis in this study represents the degree of tectonic activity for both the north-eastern and southwestern Honshu islands. The resulting principal component scores were then applied to a cluster analysis to conduct quantitative classifications, and the result provided three and nine seismotectonic provinces in the northeastern and southwestern Honshu islands, respectively.

Keywords Seismotectonic province map • Principal component analysis • Cluster analysis

T. Kumamoto (✉) • M. Tsukada
Okayama University, Fact. Science, Tsushimanaka 3-1-1, Kita-ku, Okayama 700-8530, Japan
e-mail: tkuma@cc.okayama-u.ac.jp

M. Fujita
Nuclear Regulation Authority, Roppongi 1-9-9, Minato-ku, Tokyo 106-8450, Japan

© The Author(s) 2016
K. Kamae (ed.), *Earthquakes, Tsunamis and Nuclear Risks*,
DOI 10.1007/978-4-431-55822-4_2

31

2.1 Introduction

Two types of intraplate earthquake are independently considered when constructing National Seismic Hazard Maps for Japan. One is earthquakes with a specific active fault, such as the 1995 Hyogo-ken-nambu (Kobe) earthquake of M_{JMA} 7.3 (magnitude of the Japan Meteorological Agency), for which the results of tectonic landform analysis and trenching surveys are used for evaluation. The other is background earthquakes, such as the 2000 Tottori-ken Seibu earthquake of M_{JMA} 7.3 and the 2008 Iwate-Miyagi Nairiku Earthquake of M_{JMA} 7.2, for which no surface observation data are available for magnitude and frequency evaluation (HERP [1]). Instead, both the earthquake statistics model of the Gutenberg-Richter relationship for magnitude and frequency estimation and the seismotectonic province map for counting observed moderate to small-sized earthquakes for the earthquake statistics are needed for hazard assessment. The HERP (2009) referred to the tectonic province map of Kakimi et al. [2]. This map was constructed by examining the density of active faults and earthquakes, focal mechanism of earthquakes, and the general tectonic setting. The tectonic boundaries of this map are shown in Fig. 2.1. Then, the magnitude for the maximum-size background earthquake was determined by referring to the historical earthquake records, and the frequency of this earthquake was calculated from the Gutenberg-Richter relationship and instrumentally observed seismicity. However, the following unsolved problems remain for application of the tectonic province map to the National Seismic Hazard Maps for Japan:

1. The main purpose of the tectonic province map of Kakimi et al. [2] is to evaluate the maximum-size earthquake, including an earthquake with a specific fault.
2. The scale of the map is 1:2,000,000, which is too small to discuss boundary locations.
3. The regulation for setting boundaries was based mainly on the subjective analysis of experienced researchers under the reference datasets mentioned above.

Therefore, the aim of this study is to construct a seismotectonic province map by quantitative and objective methods. For this purpose, we adopted the following four datasets: observed seismicity, the distribution of active faults, the lower limit of the seismogenic layer, and the Bouguer gravity anomaly.

The statistical methods adopted in this research for the multivariate analysis are principal component analysis and cluster analysis. The spatial resolution for the analysis was set as the second-order map grid of ca. 10 by 10 km of the Geographic Survey of Japan. Then, the four datasets were compiled for each grid, and the principal component loadings were calculated for the cluster analysis. Finally, the boundaries of the tectonic province map were depicted by referring to the result of the cluster analysis.

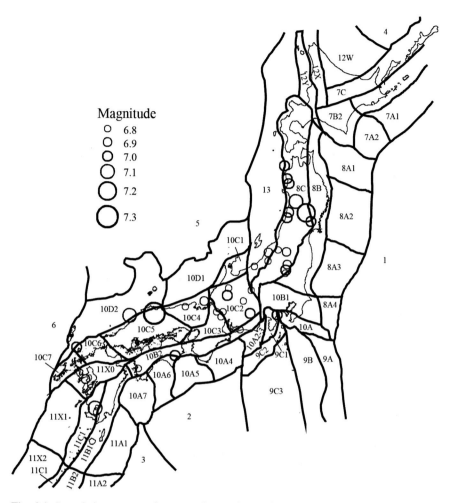

Fig. 2.1 Boundaries among seismotectonic provinces after Kakimi et al. [2]. *Circles* show intraplate blind earthquakes during 1926–2011

2.2 Data and Method

To construct the seismotectonic province map by the principal component analysis, we parameterized the observed seismicity, distribution of active faults, lower limit of the seismogenic layer, and Bouguer gravity anomaly as explained below.

2.2.1   Observed Seismicity (Fig. 2.2a)

For the instrumentally observed seismicity data, we extracted earthquakes of magnitude 4.0 or larger with depths of 20 km or shallower during 1926–1995 from the Japan Meteorological Agency's integrated hypocenter database. Next, aftershocks due to large earthquakes of magnitude 6.0 or larger in the same period were excluded by the method of the Public Works Research Institute [3]. Then, the magnitude M_{JMA} of each earthquake in a grid was converted to seismic moment Mo [dyne-cm] by the Eq. (2.1) of Takemura [4].

$$LogMo = 1.2\,M_{JMA} + 17.7 \qquad (2.1)$$

Finally, the logarithmic summation of the seismic moments of the earthquakes per year in the grid was calculated as a parameter of the observed seismicity (Fig. 2.3a).

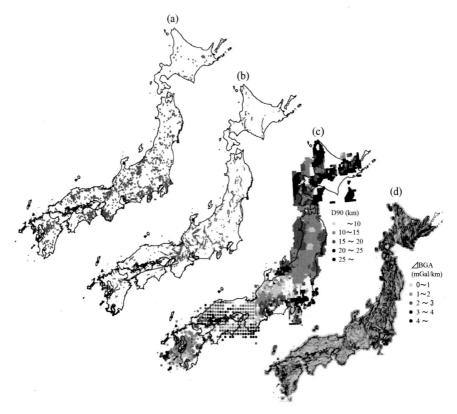

Fig. 2.2 Original parameters for the principal component analysis: (**a**) seismicity, (**b**) distribution of active faults, (**c**) lower limit depth of the seismogenic layer, and (**d**) inclination of Bouguer gravity anomaly

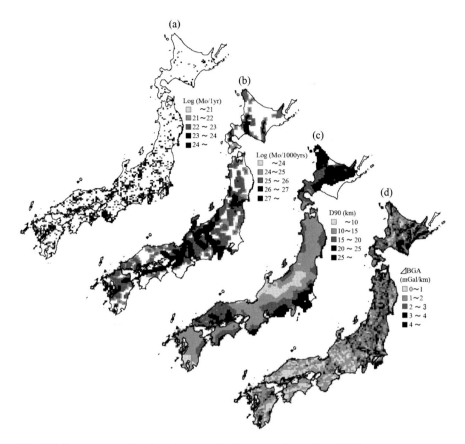

Fig. 2.3 Parameters assigned to the second-order map grid (scale 1:25,000) for the principal component analysis: (**a**) logarithmic seismic moment release rate per year, (**b**) logarithmic seismic moment release rate per 1k year, (**c**) weighted average lower limit depth of the seismogenic layer, and (**d**) averaged inclination of the Bouguer gravity anomaly

2.2.2   Distribution of Active Faults (Fig. 2.2b)

In order to include the long-term averaged seismicity data with thousands to tens of thousands of years, that is, almost identical to the average recurrence period of active faults in Japan, the distribution map of active faults was used in the principal component analysis. The seismogenic fault distribution database of Okino and Kumamoto [5], which is a recompilation of the Digital Active Fault Map of Japan (Nakata and Imaizumu, eds., [6]), was used. The length of each seismogenic fault L (km) was converted to M_{JMA} by Mastuda's Eq. (2.2) [7] and then reconverted to seismic moment Mo by equation (1).

$$\mathrm{LogL} = 0.6M_{\mathrm{JMA}} - 2.9 \qquad\qquad (2.2)$$

Next, the seismic moment Mo of each seismogenic fault was divided by the averaged recurrence interval derived from trenching surveys or empirical equations, and the seismic moment release rate per 1000 yr was calculated.

If the seismic moment release rate of a seismogenic fault is assigned to grids that are cut through by the seismogenic fault, the contrast between a grid with the seismogenic fault and a neighboring grid without the seismogenic fault becomes too large. A trial test with such contrast clarified that severe bias results in the principal component analysis. Therefore, the seismic moment release rate of such a grid was redistributed to the neighboring eight grids by using the Gaussian weighting function. This procedure corresponds to the consideration of subsidiary faults around the main traces of a seismogenic fault, which is omitted in the seismogenic fault distribution database (Okino and Kumamoto [5]). Finally, the logarithmic summation of seismic moments of earthquakes per 1000 yr in the grid was calculated for a parameter of the distribution of active faults (Fig. 2.3b).

2.2.3 Lower Limit of the Seismogenic Layer (Fig. 2.2c)

As the first order subsurface structural parameter, the lower limit of the seismogenic layer D90 was adopted in this study. The D90 shows the 90 % depth among instrumentally observed seismicity in a 0.025–0.2-degree grid (Seismotectonics Research Group ed. [8]). This lower limit of the seismogenic layer data relates to the seismogenic fault width and shows locality in Japan. Interpolation with the weight of the reciprocal of a distance to a grid was applied to the D90 data, and the parameter for the principal component analysis was created (Fig. 2.3c).

2.2.4 Bouguer Gravity Anomaly (Fig. 2.2d)

Bouguer gravity anomaly data are corrected to indicate subsurface density structure, and many reports show that the specific large inclination of the Bouguer gravity anomaly corresponds to the distribution of active faults (e.g., Hagiwara ed. [9]). Thus, the data of the assumed density of 2.67 g/cm^3 with 1 km resolution from the Gravity CD-ROM of Japan Ver.2 (Geological Survey of Japan [10]) were adopted in this study for the second subsurface structural parameter. The average inclination in a grid was calculated for the principal component analysis (Fig. 2.3d).

Because the unit of each parameter was different, the correlation matrix method was used in the principal component analysis. As a result, principal component loadings, eigenvalues, coefficients of determination, cumulative coefficients of determination, and principal component scores were calculated for four parameters.

2.3   Result and Discussion

Figure 2.4 shows the first, second, third, and fourth principal component loadings for the four parameters of observed seismicity, distribution of active faults, lower limit of the seismogenic layer, and Bouguer gravity anomaly in (a) the northeastern Honshu, (b) southwestern Honshu, (c) Kyushu, (d) Hokkaido, and (e) Kanto districts resulting from the principal component analysis.

The first principal component loadings (F1) in Fig. 2.4a of northeastern Honshu showing a 34 % proportion indicate that the observed seismicity, distribution of active faults, and Bouguer gravity anomaly parameters have positive values and that the lower limit of the seismogenic layer has a negative value. This result means that the short-term observed seismicity matches well the estimated long-term

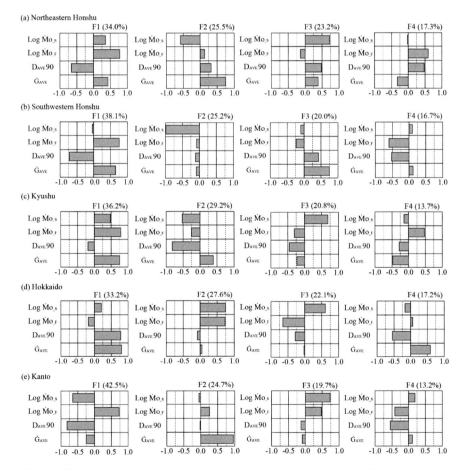

Fig. 2.4 The first to fourth principal component loadings for seismicity, active faults, lower limit depth of the seismogenic layer, and inclination of the Bouguer gravity anomaly of each region

averaged seismicity from active faults. In addition, those areas where the density of the observed seismicity and distribution of active faults are both high show a relatively thin seismogenic layer and complex subsurface structure deduced from the Bouguer gravity anomaly. Thus, we consider that the first principal component might indicate the degree of tectonic activity.

The results of the relations between the principal component loadings and four observed parameters of (b) southwestern Honshu and (c) Kyushu in Fig. 2.4 show a similar tendency. Thus, we expect that the spatial distribution of the first principal component loading relates to the seismotectonic provinces and adopted it as a parameter for the following cluster analysis. However, to the contrary, the first principal component loading (F1) in Fig. 2.4d of Hokkaido showing a 33 % proportion indicates that the lower limit of the seismogenic layer and the Bouguer gravity anomaly parameters are large positive values and that the remaining parameters are relatively small. Additionally, the first principal component loadings (F1) in Fig. 2.4e of Kanto indicate a different tendency from the other districts, showing that the first principal component loadings of the observed seismicity and Bouguer gravity anomaly parameters are largely negative. The reason for the different results for the Hokkaido and Kanto districts might relate to the distance between the district and plate boundary axis. The closer distance of the Hokkaido and Kanto districts results in complexity of the deep part of the tectonic structure, such as the depth and the shape of the subducting oceanic plate and the related seismicity and gravity anomaly.

Hereafter, cluster analysis is applied to the first principal component scores of northeastern Honshu and southwestern Honshu, respectively, from the viewpoint of the tectonic meaning of intraplate shallow earthquakes and active faults as judged from the principal component loading results (Fig. 2.4). The statistical distances among grids in each district were measured by the group average method, and clusters were calculated in similarity order. The number of clusters is after that of Kakimi et al. [2], and four and six were set for northeastern Honshu and southwestern Honshu, respectively (Fig. 2.5). The result shows that each cluster in both northeastern and southwestern Honshu distributes with strong spatial relation, although no parameter regarding contiguity was involved in the principal component analysis. Thus, the similarity of the first principal component score based on the objective index of statistical distance is considered to have usefulness for considering the spatial correlation of tectonic provinces.

Then, the tectonic province boundaries were set at the cluster boundaries between areas of clusters with the same category, as shown in Fig. 2.6a for northeastern Honshu and Fig. 2.6b for southwestern Honshu. The exception was the case of isolated cluster patches consisting of 10 or fewer grids, which were subjectively determined to be ignored or incorporated into a neighboring cluster by the shape and size of each patch. The following describes the distinctive features of our results (Fig. 2.6) and the tectonic province map of Kakimi et al. [2] (Fig. 2.1).

Northeastern Honshu was divided into two large seismotectonic provinces by Kakimi et al. [2]. This boundary is located at the eastern foot of the Ou-Backbone Mountain Range and partly overlaps with west-dipping reverse fault systems, such

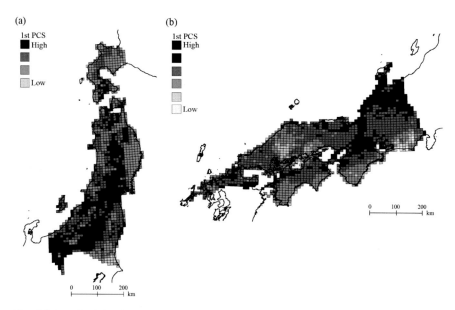

Fig. 2.5 Result of the cluster analysis using the first principal component loadings. Four and six clusters are shown for northeastern and southwestern Honshu, respectively

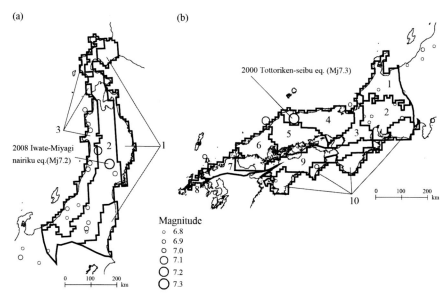

Fig. 2.6 Boundaries among the seismotectonic provinces defined in this study: (**a**) northeastern Honshu island with three provinces and (**b**) southwestern Honshu island with nine provinces

as the Kitakami lowland west boundary fault system and the Fukushima basin west boundary fault system. To the contrary, our result (Fig. 2.6a) shows that northeastern Honshu is divided into three large provinces. The eastern boundary is located 10–50 km eastward of the eastern foot of the Ou-Backbone Mountain Range, and the western boundary is located 10–50 km westward of the western foot of the Ou-Backbone Mountain Range, where the Yokote basin east boundary fault system and other east-dipping reverse fault systems are identified. Therefore, the two parallel boundaries in this study involve the Ou-Backbone Mountain Range and separate from both outer provinces with relatively low seismicity. This result is consistent with the idea that the Ou-Backbone Mountain Range was uplifted by the conjugate reverse fault system beneath the mountain ranges. From the viewpoint of our result, the background of the 2008 Iwate-Miyagi Nairiku Earthquake is then located in a province with high density of seismicity and active faults.

In this study, southwestern Honshu was divided into nine seismotectonic provinces (Fig. 2.6b), and the number of provinces is the same as that of Kakimi et al. [2]. There are three major differences in southwestern Honshu between Fig. 2.6b and Fig. 2.1: First, in Kakimi et al. [2], the Median Tectonic Line (MTL), the longest and one of the most active fault systems in Japan, is excluded from the province as an exception of the specific fault and overlaps the boundary between the outer and inner zones of southwestern Honshu. However, our objective calculation results in a province involving the MTL without exceptional consideration, influenced mainly by the density contrast of the active fault distribution (No. 9 of Fig. 2.6b). Second, in Kakimi et al. [2], the seismotectonic boundary between the Kinki district with a high density of seismicity and active faults and the Chugoku district with a low density of each is located at the Yamazaki fault system and its extensions. This boundary excludes the background earthquake of the 2000 Tottori-ken Seibu Earthquake from the province with a high density of seismicity and active faults. To the contrary, our result (Fig. 2.6b) shows that the boundary is located westward by 20–30 km in the southern section and 50–80 km in the northern section. Thus, both the source fault of the 2000 Tottori-ken Seibu Earthquake and the Shimane nuclear power plant site in the Shimane Peninsula are included in the same seismotectonic province of the Kinki district where the high density of seismicity and active faults are observed. Third, the Chugoku district is divided into two seismotectonic provinces corresponding to the density difference of active fault distribution (Fig. 2.3b, whereas it is one large province in the map of Kakimi et al. [2].

2.4 Future Challenges

In this study, a new quantitative and objective seismotectonic province map for improvement of background earthquake assessment was constructed by combining the four datasets of observed seismicity, distribution of active faults, lower limit of the seismogenic layer, and Bouguer gravity anomaly and the statistical methods of principal component analysis and cluster analysis. The resulting map in Fig. 2.6 shows different tectonic province boundaries and the necessity of a different explanation regarding the 2000 Tottori-ken Seibu Earthquake and the 2008 Iwate-Miyagi Nairiku Earthquake, which represent large background earthquakes. Future improvement is still needed in regard to ways to estimate the largest magnitude and associated frequency of background earthquake in a rational and objective seismic hazard assessment, not only in each province but also at the site of a nuclear power plant.

Acknowledgement This work is supported by Japan Nuclear Energy Safety Organization (2008–2011) and partly Grant-in-Aid for Scientific Research (21510191).

References

1. HERP (2009) Technical report for National Seismic Hazard Maps for Japan Part3 (in Japanese). http://www.jishin.go.jp/main/chousa/09_yosokuchizu/g_kakuritsuron.pdf
2. Kakimi T et al (2003) A seismotectonic province map in and around the Japanese islands (in Japanese). J Seismol Soc Jpn 55:389–406
3. Public Works Research Institute (1983) Investigation of frequency and magnitude of fore-shocks and aftershocks (in Japanese). Technical Note of PWRI: No.1995, Tsukuba
4. Takemura M (1998) Scaling law for Japanese intraplate earthquakes in special relations to the surface faults and damages (in Japanese). J Seismol Soc Jpn 51:211–228
5. Okino N, Kumamoto T (2007) New integrated linear source model for intraplate earthquakes based on digitized databases (in Japanese). Active Fault Res 27:75–94
6. Nakata T, Imaizumi T (eds) (2002) Digital active fault map of Japan (in Japanese). University of Tokyo Press, Tokyo
7. Matsuda T (1975) Magnitude and recurrence interval of earthquakes from a fault (in Japanese). J Seismol Soc Jpn 28:269–283
8. Seismotectonics Research Group (ed) (2005) Upper and lower limit of the seismogenic layer in Japanese provinces – for improvement of fault width estimation of intraplate earthquake – (in Japanese). Association for the Development of Earthquake Prediction, 102p, Tokyo
9. Hagiwara T (1991) Earthquakes of Japanese Islands. Kashima Press, 215p, Tokyo
10. Geological Survey of Japan (2004) Gravity CD-ROM of Japan, Ver.2 (in Japanese), GSJ, Tsukuba

Chapter 3
Multiple Regression Analysis for Estimating Earthquake Magnitude as a Function of Fault Length and Recurrence Interval

Takashi Kumamoto, Kozo Oonishi, Yoko Futagami, and Mark W. Stirling

Abstract Multiple regressions are developed using world earthquake data and active fault data, and the regressions are then evaluated with Akaike's Information Criterion (IEEE Trans Autom Control, 19(6):716–723). The AIC method enables selection of the regression formula with the best fit while taking into consideration the number of parameters. By using parameters relevant to earthquakes and active faults in the regression analyses, we develop a new empirical equation for magnitude estimation as Mw = 1.13logLs + 0.16logR + 4.62.

Keywords Multiple regression analysis • Magnitude • Fault length • Recurrence interval

3.1 Introduction

Many empirical equations for estimating earthquake magnitude have been developed in Japan. The most famous of these is the so-called Matsuda's Equation [2], which is widely used for constructing seismic hazard maps of Japan (e.g., The Headquarters for Earthquake Research Promotion [3]). Matsuda's Equation (Eq. 3.1) below is based on 14 earthquakes dated from the 1891 Nobi earthquake (M_{JMA} 8.0: Japan Meteorological Agency magnitude) to the 1970 Southeastern Akita Prefecture earthquake (M_{JMA} 6.2).

T. Kumamoto (✉)
Faculty of Science, Okayama University, Tsushimanaka 3-1-1, Kita-ku, Okayama 700-8530, Japan
e-mail: tkuma@cc.okayama-u.ac.jp

K. Oonishi
Shikoku Electric Power CO., Inc, Marunouchi 2-5, Takamatsu 760-8573, Japan

Y. Futagami
National Research Institute for Cultural Properties, Tokyo, Ueno Park 13-43, Taito-ku, Tokyo 110-8713, Japan

M.W. Stirling
GNS Science, 1 Fairway Drive, Avalon 5010, New Zealand

© The Author(s) 2016
K. Kamae (ed.), *Earthquakes, Tsunamis and Nuclear Risks*,
DOI 10.1007/978-4-431-55822-4_3

43

$$\log L = 0.6M_{JMA} - 2.9 \qquad (3.1)$$

In Matsuda's Equation, the fault parameter length L might be regarded as sub-surface. On the contrary, the length of an active fault recognized on the surface from tectonic geomorphology must be used for magnitude estimation prior to the occurrence of the next earthquake. However, a problem occurs in estimating magnitudes of isolated active faults with lengths significantly shorter than the thickness of the seismogenic layer, which is estimated to be roughly 15–20 km in Japan. Thus, another empirical equation for estimating earthquake size is the following empirical equation between seismic moment Mo and fault area by Irikura and Miyake [4] as:

$$
\begin{aligned}
S &= 2.23 \times 10^{-15} Mo^{2/3} \text{ for } Mo < 7.5 \times 10^{25} \text{dyne-cm} \\
S &= 4.24 \times 10^{-11} Mo^{1/2} \text{ for } Mo \geqq 7.5 \times 10^{25} \text{dyne-cm}
\end{aligned} \qquad (3.2)
$$

This equation is also widely used especially in strong ground motion prediction in Japan.

However, the question of large uncertainties in the earthquake scaling relation remains in both equations. To resolve the above issues, multiple regressions are developed using world earthquake data and active fault data, and the regressions are then evaluated with AIC (Akaike's Information Criterion, Akaike, 1974). The AIC method enables selection of the regression formula with the best fit while taking into consideration the number of parameters. By using parameters relevant to earthquakes and active faults such as stress drop, average slip rate, and recurrence interval, in the regression analysis, we develop a new empirical equation for magnitude estimation.

3.2 Data

The database used in this paper is compiled from the intraplate earthquake datasets listed below.

1. Earthquake and active fault data of Wells and Coppersmith [5]:

These data were compiled to develop empirical relationships between earthquake magnitude and various fault parameters. The data are for the years 1857–1994, and the number of data items is 244. The data fields include earthquake location, name, date, slip type, magnitude (surface wave magnitude Ms, moment magnitude Mw, seismic moment Mo), subsurface and surface rupture length, fault width, fault area, and maximum/average surface slip amount. Values thought to have low reliability are given in parentheses.

2. Earthquake data of Anderson et al. [6]:

These historical earthquake data were used to estimate earthquake magnitude as a function of the length and slip rate of the causative fault. The earthquake data were collected for the time period 1811–1994, and the number is 43 in total. The data fields include location, Mo, Mw, fault length, and slip rate.

3. Earthquake and active fault data of Mohammadioun and Serva [7]:

The characteristic aspect of this dataset is a list of variations in stress drop for different slip types, including strike-slip, normal, or reverse faults in the 1857–1994 earthquakes worldwide, plus the Umbria earthquake in 1998 (Ms 5.7), the Chi-Chi earthquake in 1999 (Ms 7.6), and the Izmit earthquake in 1999 (Ms 7.4). The number of earthquakes in the dataset is 90, and the data fields include maximum surface slip amount, static stress drop $\Delta\sigma_1$, dynamic stress drop $\Delta\sigma_2$, and the ratio of dynamic stress drop to static stress drop.

4. Earthquake and active fault data of Stirling et al. [8]:

The number of data items is 389, and several scaling laws are developed in this paper to compare instrumental and pre-instrumental data. The data fields include slip type, magnitude (Ms, M_{JMA}, Mw, and Mo), minimum/maximum seismogenic fault length, minimum/maximum surface rupture length, minimum/maximum fault width, and maximum/average surface slip amount.

3.3 Parameters for Analysis

We first examine whether the strength of asperities on the fault plane can be quantified by the stress drop $\Delta\sigma$. Cotton et al. [9] showed the importance of $\Delta\sigma$ for strong ground motion and also showed large variability of $\Delta\sigma$. The 90 data items for static stress drop $\Delta\sigma_1$ and dynamic stress drop $\Delta\sigma_2$ compiled from [7] are used here as stress drop $\Delta\sigma$. The static stress drop $\Delta\sigma_1$ is a value calculated from the average slip amount D_{ave} and coseismic rupture length L from geological/seismological observations with the following equation:

$$\Delta\sigma_1 \propto \frac{D_{ave}}{L} \qquad (3.3)$$

The dynamic stress drop $\Delta\sigma_2$ is a value from the spectrum of the seismic wave record.

The recurrence interval R of a fault is estimated directly from the observed displacement of layers and the dating of layers on the historical earthquake record or trench excavation results. However, this estimation is not always possible, and calculated estimation is conducted by dividing the average slip amount on the fault by the average slip rate S_{ave} as the following formula.

$$R = \frac{D_{ave}}{S_{ave}} \tag{3.4}$$

The average slip rate S_{ave} is also an indirect value derived from the cumulative slip amount D divided by the dating of layers/tectonic landforms T for which that slip amount was obtained.

$$S_{ave} = \frac{D}{T} \tag{3.5}$$

The average recurrence interval R is calculated for the average slip rate S_{ave} by using the dataset of 43 earthquakes in this research (Table 3.1). However, it is difficult to estimate the average slip rate S_{ave} in Eq. (3.5) because the cumulative slip amount D is not constant along fault traces. Thus, uncertainties of slip rate might be large, and the simple mean between S_{min} and S_{max} leads to an underestimation or overestimation which is inappropriate for magnitude estimation. Therefore, just as in Anderson et al. (1996), 100 random numbers were generated in the range between S_{min} and S_{max}, and the distribution of the average slip rate S_{ave} was determined in Table 3.1 for calculation of R in Eq. (3.4).

3.4 Results and Discussion

Large variances of earthquake magnitude to the same surface rupture length Ls are observed in Fig. 3.1 from [8]. Then Fig. 3.2 shows the relationship between earthquake magnitude and surface rupture length from our compiled dataset in Table 3.1. In Fig. 3.2, the dynamic stress drop $\Delta\sigma_2$ was divided with four marks: 50 bars or less, 50–70 bars, 70–90 bars, and 90 bars or more. According to Fig. 3.2, the stress drop is almost always 50 bars or less for earthquakes the fault length for which exceeds 100 km, though the data contain many values of 90 bars or more for fault lengths of 20 km or shorter. For fault lengths between 20 km and 100 km, in particular, near the fault length of 40 km which corresponds to an aspect ratio of two seismogenic-layer earthquakes with a large stress drop display large magnitudes, and earthquakes with a small stress drop exhibit a relatively small magnitude, even if the surface rupture length is the same. Therefore, it is possible to infer that the dynamic stress drop $\Delta\sigma_2$ could be an additional parameter for estimating earthquake magnitude.

Then, a total of six variables were set for regression analysis: surface rupture length Ls, seismogenic fault length L_{sub}, maximum slip amount D_{max}, average slip amount D_{ave}, static stress drop $\Delta\sigma_1$, and dynamic stress drop $\Delta\sigma_2$. Single and multiple regression analyses were conducted to estimate moment magnitude Mw, and the goodness of fit arising from varying combinations of one variable, two variables and three variables were evaluated with AIC, the value of the coefficient of correlation for single regression analysis and the values of coefficient of

Table 3.1 Data for multiple regression analysis compiled from [5–8]

Event	Date	Lat	Lon	Type	Mw	Ls	D_{ave}	S_{min}	S_{max}	S_{ave}	$\Delta\sigma1$	$\Delta\sigma2$
New Madrid	1811/12/16	36.6	−89.6		8.2	95	6	0.01	2	1.005		
Marlborough	1848/10/15	−40.6	174.5	S	7.5			4	10	7		
Fort Tejon	1857/1/09	34.9	−119.1	S	7.7	297	5.53	16	43	29.5	9.5	18.8
Hayward	1868/10/21	37.5	−122	S	6.8	52		8	10	9	5.63	21.7
Owens Valley	1872/3/26	36.7	−118.1	SN	7.6	108	6	1	3	2	30.6	51.2
Canterbury	1888/9/1	−42.5	172.6	S	7.2	35	2.5	11	25	18		
Nobi	1891/10/27	35.6	136.6	S	7.4	80	4.2	1	10	5.5	30	80.4
Rikuu/Senya	1896/8/31	39.5	140.7	R	7.2		2.59	0.1	1	0.55	33	57
San Francisco	1906/4/18	37.8	−122.6	S	7.9	432	3.9	15	28	21.5	4.24	1.13
Pleasant Valley	1915/10/3	40.3	−117.6	N	7.2	62	3.5	0.3	1	0.65	28.1	59
Tango	1927/3/7	35.5	135.2	SR	6.8	14	2.9	0.01	1	0.505	64.3	657
North Izu	1930/11/25	35.1	139.1	SR	6.9	35	2.95	1	10	5.5	32.6	83
Long Beach	1933/3/11	33.6	−118	S	6.4		0.2	0.1	6	3.05		
Parkfield	1934/6/7	35.8	−120.4	S	6.1	20		29	39	34		
Erzincan	1939/12/25	38	40.3	S	7.7	360	2.43	5	25	15	6.25	5.92
Imperial Valley	1940/5/19	32.8	−115.5	S	6.9	60	1.83	18	23	20.5	29.5	31
Erbaa	1942/12/20	40.7	36.5	SN	6.7	47	0.59	5	25	15	12.8	45
Bolu	1944/2/1	40.9	32.6	S	7.5	180	2.28	5	25	15	6	10
Kem Country	1952/7/21	35.3	−118.7	RS	7.4	57	2.6	3	8.5	5.75	15.8	80
Canakkale	1953/3/18	39.9	27.4	S	7.2	58	2.45	5	25	15	22.5	32.6
Fairview Peak	1954/12/16	39.2	−118.2	SN	7.2	57	2.45	0.01	1	0.505	21.6	33.5
Dixie Valley	1954/12/16	39.6	−118.2	SN	6.9	45	1.8	0.3	1	0.65	25.3	23.9
San Miguel	1956/2/9	31.7	−115.9	SR	6.6	22	0.5	0.1	0.5	0.3	12.3	83
Hebgen Lake	1959/5/15	44.8	−111.2	N	7.3	26.5	2.47	0.8	2.5	1.65	69	212
Niigata	1964/6/18	38.4	139.2	R	7.6		3.3	0.01	1	0.505		

(continued)

Table 3.1 (continued)

Event	Date	Lat	Lon	Type	Mw	Ls	Dave	S_{min}	S_{max}	S_{ave}	$\Delta\sigma1$	$\Delta\sigma2$
Parkfield	1966/6/28	35.8	−120.4	S	6.3	38.5	0.38	29	39	34	1.56	15.1
Mudurna Valley	1967/7/22	40.7	31.2	S	7.3	80	1.35	5	25	15	9.75	28.4
Borrego Mountain	1968/4/9	33.2	−116.1	S	6.8	31	0.54	1.4	5	3.2	3.68	41.9
San Fernando	1971/2/9	34.3	−118.4	FG	5.6	16	1.48	2	7.5	4.75	47	67.4
Luhuo	1973/2/8	31.3	100.7	S	7.5	89	1.3	5	10	7.5	12.1	20.4
Coyote Lake	1979/8/6	37	−121.5	S	5.8	14.4		15	19	17	3.13	19.8
El Centro	1979/10/15	32.8	−115.4	S	6.5	30.5	0.8	18	23	20.5	7.87	36.1
Daolu	1981/1/23	31	101.2	S	6.6	44		5	10	7.5	10.2	24.7
Coalinga	1983/5/2	36.3	−120.5	RS	6.4			1	7	4		
Borah Peak	1983/10/28	44.2	−113.8	NS	6.9	34	1.15	0.07	0.3	0.185	23.8	87
Morgan Hill	1984/4/24	37.32	−121.7	S	6.3			3	6.4	4.7		
No Palm Springs	1986/7/8	33.9	−116.6	SR	6.1	9		14	25	19.5		
Edgecumbe	1987/3/2	−38	176.5	N	6.5	14	1.7	1.3	2.8	2.05	48.3	67
Superstition Hills	1987/11/24	33	−115.8	S	6.6	27	0.54	2	6	4	10.2	36.5
Loma Prieta	1989/10/18	37.2	−121.9	SR	6.9			12	28	20		
Luzon	1990/7/16	15.7	121.1	S	7.7	110		10	20	15	15.5	31
Landers	1992/6/28	34.2	−116.4	S	7.3	71	2.95	0.08	2	1.04	25.3	48.1
Northridge	1994/1/17	34.2	−118.5	R	6.7			1.4	1.7	1.55		

Fig. 3.1 Relationship
between earthquake
magnitude and surface
rupture length in [8]

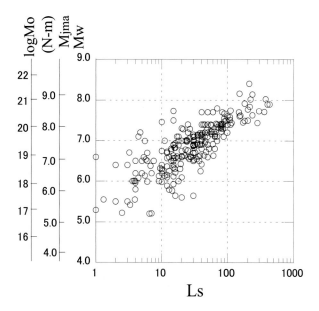

Fig. 3.2 Relationship
between earthquake
magnitude and surface
rupture length with different
marks according to stress
drop in Table 3.1

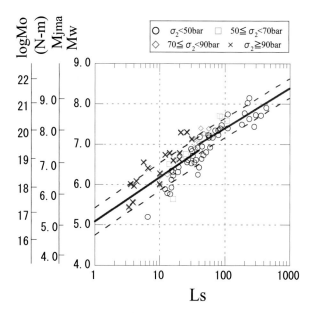

determination (Table 3.2). Among various combinations, the Akaike's Information
Criterion [1] value was minimized by taking account of the two variables, surface
rupture length Ls, and dynamic stress drop $\Delta\sigma_2$ as Eq. (3.6), though the coefficient
of correlation between Ls and $\Delta\sigma_2$ shows 0.69, which means that some multi-
collinearity effects are assumed.

Table 3.2 Comparison of goodness of fit from single and multiple regression analyses

Response variable	Explanatory variable	AIC	Coef.	
Mw	$\log\Delta\sigma_1$	168.00	0.12	
	$\log\Delta\sigma_2$	160.00	0.04	

Response variable	Explanatory variable	AIC	Coef.	Coef. in variables
Mw	$\log Ls$-$\log\Delta\sigma_2$	−26.16	0.90	0.69
	$\log L_{sub}$-$\log\Delta\sigma_2$	29.65	0.78	0.36
	D_{max}-$\log\Delta\sigma_2$	96.00	0.59	0.17
	D_{ave}-$\log\Delta\sigma_2$	61.32	0.64	0.18
	$\log Ls$-$\log\Delta\sigma_1$	28.34	0.81	0.05
	$\log L_{sub}$-$\log\Delta\sigma_1$	29.05	0.78	0.18
	D_{max}-$\log\Delta\sigma_1$	102.53	0.56	0.43
	D_{ave}-$\log\Delta\sigma_1$	68.36	0.60	0.23
	$\log Ls$-D_{max}-$\log\Delta\sigma_2$	−26.08	0.90	0.41, 0.69, 0.17
	$\log Ls$-D_{ave}-$\log\Delta\sigma_2$	−22.15	0.90	0.66, −0.72, −0.18
	$\log L_{sub}$-D_{max}-$\log\Delta\sigma_2$	18.17	0.82	0.67, −0.36, −0.15
	$\log L_{sub}$-D_{ave}-$\log\Delta\sigma_2$	13.56	0.80	0.66, −0.44, −0.11
	$\log Ls$-D_{max}-$\log\Delta\sigma_1$	22.73	0.83	0.64, 0.05, 0.41
	$\log Ls$-D_{ave}-$\log\Delta\sigma_1$	5.68	0.84	0.66, −0.26, 0.22
	$\log L_{sub}$-D_{max}-$\log\Delta\sigma_1$	19.99	0.82	0.67, 0.18, 0.48
	$\log L_{sub}$-D_{ave}-$\log\Delta\sigma_1$	13.59	0.80	0.66, −0.08, 0.29

$$Mw = 1.62\log Ls + 0.76\Delta\sigma_2 + 3.07 \qquad (3.6)$$

However, the dynamic stress drop $\Delta\sigma_2$ is a value obtained by the spectrum of the seismic wave record observed after the occurrence of an earthquake. Therefore, it is an inappropriate parameter for the estimation of future earthquakes. Our alternative approach is to find a proxy parameter of dynamic stress drop $\Delta\sigma_2$. Figures 3.3 and 3.4 show the relationship between slip rate and $\Delta\sigma_2$ and between recurrence interval and $\Delta\sigma_2$, respectively. Figure 3.3 shows that the dynamic stress drop $\Delta\sigma_2$ is large when the average slip rate S is small and the dynamic stress drop $\Delta\sigma_2$ becomes small when the average slip rate S is large. One problem arising when slip rate is used for prediction of earthquake magnitude is that the average slip rate S has a small value of 0.01–1 mm/year in the active fault catalog, and most of them that are calculated from Eq. (3.5) need age and displacement data derived from tectonic landforms in field surveys. On the other hand, the average recurrence interval R could be derived in both Eq. (3.4) and trench excavations in field surveys. Figure 3.4 shows a relationship between recurrence interval and dynamic stress drop $\Delta\sigma_2$ in which longer recurrence intervals correlated with larger stress drops.

We next conducted regression analyses with the average recurrence interval R determined from average displacement D_{ave} and average slip rate in Table 3.1. Moment magnitude Mw was set as the response variable, and surface rupture length Ls and average recurrence interval R were set as the explanatory variables. The

Fig. 3.3 Relationship
between average slip rate
and the dynamic stress drop
$\Delta\sigma2$ in Table 3.1

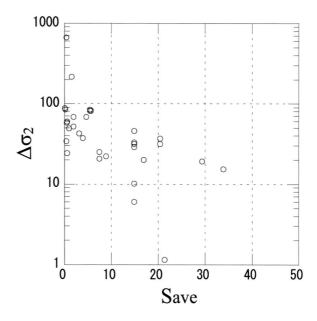

Fig. 3.4 Relationship
between recurrence interval
and the dynamic stress drop
$\Delta\sigma2$ in Table 3.1

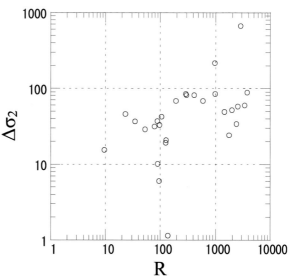

following regression equation was obtained based on the average and standard
deviation for each regression coefficient:

$$Mw = 1.13\log Ls + 0.16\Delta\sigma_2 + 4.62 \qquad (3.7)$$

This Eq. (3.7) is then compared to Matsuda's Equation (Eq. 3.1) in Fig. 3.5.
Equation (3.7) includes the average recurrence interval R in its explanatory

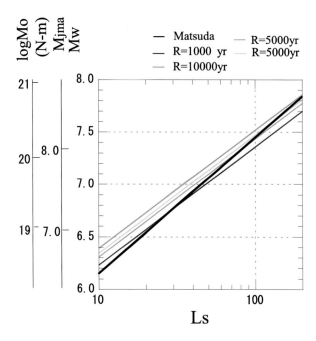

Fig. 3.5 Comparison of the multiple regression equation in this study with different recurrence intervals and Matsuda's Equation

variables, and thus 1000, 3000, 5000, and 10,000 years are given as typical examples of the average recurrence intervals in Fig. 3.5. For surface rupture length Ls of 20 km, we find that the magnitude obtained with Eq. (3.7) is larger than Matsuda's Equation, for all recurrence interval estimates. In contrast, the magnitude is smaller than that of Matsuda's Equation when Ls is 80 km for average recurrence intervals of less than 3000 years and larger than Matsuda's Equation for average recurrence intervals more than 3000 years. Furthermore, the magnitude is smaller than Matsuda's Equation for an average recurrence interval of 1000 years or less. A difference of 0.2 in magnitude corresponds to a difference of 1000 and 10,000 years in recurrence interval. Therefore, earthquakes of different magnitude in faults of similar lengths can be explained by different recurrence intervals.

3.5 Summary and Conclusion

We have developed new regressions for estimating earthquake magnitude from the fault parameters such as stress drop and recurrence interval. The resulting equation was obtained. The equation shows that for a fault possessing an average recurrence interval of 1000 years or less and a length of 30 km or more, the magnitude estimated from our equation is less than that produced by Matsuda's Equation. Conversely, the magnitude is larger than that of Matsuda for a fault length of 80 km and an average recurrence interval of 3000 years or more. A difference of 0.2 in magnitude between average recurrence intervals of 1000 years and 10,000 years was also shown.

Acknowledgment The authors thank Ms. Miki Tachibana for data collection. This work is supported by Japan Nuclear Energy Safety Organization (2008–2011) and partly Grant-in-Aid for Scientific Research (18540423).

References

1. Akaike H (1974) A new look at the statistical model identification. IEEE Trans Autom Control 19(6):716–723
2. Matsuda T (1975) Magnitude and recurrence interval of earthquakes from a fault (in Japanese). J Seismol Soc Japan 28:269–283
3. The Headquarters for Earthquake Research Promotion (2009) A recipe proposed for estimating strong ground motions from specific earthquakes (in Japanese). http://www.jishin.go.jp/main/chousa/09_yosokuchizu/g_furoku3.pdf
4. Irikura K, Miyake H (2001) Prediction of strong ground motion for scenario earthquakes (in Japanese). J Geogr 110:849–875
5. Wells DL, Coppersmith KJ (1994) New empirical relationships among magnitude, rupture length, rupture area, and surface displacement. Bull Seismol Soc Am 84:974–1002
6. Anderson JG et al (1996) Earthquake size as a function of fault slip rate. Bull Seismol Soc Am 86:683–690
7. Mohammadioun B, Serva L (2001) Stress drop, slip type, earthquake magnitude, and seismic hazard. Bull Seismol Soc Am 91:694–707
8. Stirling MW et al (2002) Comparison of earthquake scaling relations derived from data of the instrumental and pre-instrumental era. Bull Seismol Soc Am 92:812–830
9. Cotton et al (2013) What is sigma of the stress drop? Seismol Res Lett 84:42–48

Chapter 4
Coseismic Tsunami Simulation Assuming the Displacement of High-Angle Branching Active Faults Identified on the Continental Slope Around the Japan Trench

Shota Muroi and Takashi Kumamoto

Abstract The aim of this study is to demonstrate the tsunami potential caused by high-angle branching faults with relatively low net slip compared to that of the 2011 off the Pacific coast of Tohoku (Tohoku-oki) earthquake of Mw9.0, located in the upper part of the mega-thrust along the Japan Trench where the Tohoku-oki earthquake ruptured, as deduced from the distribution of active faults illustrated by a bathymetric geomorphological study and seismic profile records (Nakata et al. Active faults along Japan Trench and source faults of large earthquakes. http://www.jaee.gr.jp/event/seminar2012/eqsympo/pdf/papers/34.pdf. 19 Dec 2012). The results show that the expected tsunami from high-angle branching faults becomes about one and a half times as high as the case of low-angle thrust faults. This demonstrates the importance of the distribution of high-angle branching faults on the continental slope and their subsurface structure in tsunami hazard assessment.

Keywords Japan Trench • High-angle branching faults • Tsunami potential

4.1 Introduction

The recent development of a detailed digital bathymetry dataset in Japan enables the interpretation of offshore tectonic landforms and the identification of the distribution of active faults, especially around plate boundaries, such as the Japan Trench and the Nankai Trough. For example, Nakata et al. [1] reported several newly identified tectonic landforms and active faults on the continental slope situated parallel to the axis of the Japan Trench where the 2011 off the Pacific

S. Muroi
Kokusai Kogyo CO., LTD, Nishinagasumachi 1-1-15, Amagasaki 660-0805, Japan

T. Kumamoto (✉)
Faculty of Science, Okayama University, Tsushimanaka 3-1-1, Kita-ku, Okayama 700-8530, Japan
e-mail: tkuma@cc.okayama-u.ac.jp

© The Author(s) 2016   55
K. Kamae (ed.), *Earthquakes, Tsunamis and Nuclear Risks*,
DOI 10.1007/978-4-431-55822-4_4

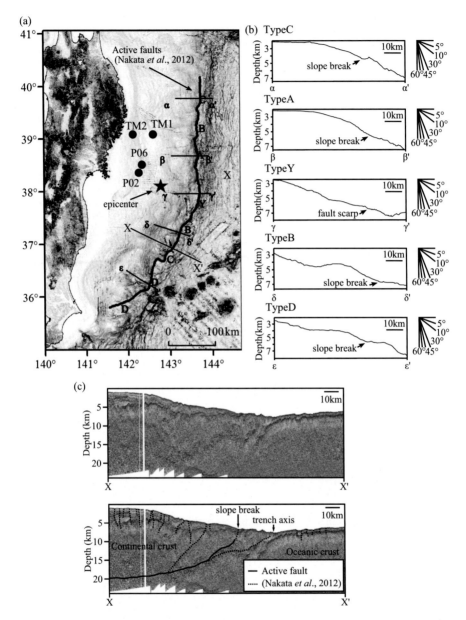

Fig. 4.1 Tectonic landforms identified off the coast of Tohoku. (**a**) The submarine active fault distributions of Nakata et al. (2012), (**b**) typical tectonic landform profiles of the sections in (**a**), (**c**) multichannel seismic profile record (top, Kato [3]; bottom, our interpretation)

coast of Tohoku (Tohoku-oki) Earthquake of Mw9.0 ruptured, by using 3D-anaglyph figures derived from a 150-m resolution Digital Bathymetry Model.

The present study demonstrates (a) the tsunami potential produced by a future earthquake due to active faults on the continental slope identified in the fault distribution map of Nakata et al. [1] (Fig. 4.1a, hereafter, Nakata map) and (b) the influence of high-angle branching faults. Although there is a lack of consensus regarding the type of tectonic landforms and active faults on the continental slope along the Japan Trench from the interpretation of seismic profile records (e.g., Okamura [2]), we discuss the tsunami potential based on the Nakata map produced from geomorphological studies because the existing planar subsurface structural dataset is still insufficient to provide an unequivocal representation of these continental slope features.

4.2   Active Fault Distribution

The earthquake source fault model in this study is constructed from the active faults on the continental slope along the Japan Trench displayed in the Nakata map with a total length of more than 400 km. These active faults can be roughly divided into the following two groups: The first group consists of faults situated at the base of the continental slope along the axis of the Japan Trench offshore between Kuji, Iwate Prefecture (40°N) and Mito, Ibaraki Prefecture (36°N) displaying a relatively linear form along their strike with fresh scarps indicating recent activity (Fig. 4.1b, sections of type Y). The second group consists of faults situated on the continental slope continuing outward on both sides of the first group faults that display a relatively sinuous form with slope breakpoints (Fig. 4.1b, sections of type A, B, C, and D).

The second group active faults in the Nakata map can be recognized in the seismic profiling record of Kato [3]. The upper figure of Fig. 4.1c shows an example of a seismic profile record along the X-X′ traverse line in Fig. 4.1a, and the lower figure of Fig. 4.1c shows our interpretation of type C high-angle branching faults with surface slope break on the continental slope. In this seismic profile record, the type C fault branches from the decollement plane between the continental and oceanic crusts with a relatively high angle of 25–28° compared to the low-angle decollement plane of approximately 14°. Ueta and Tani [4] performed an experimental simulation of thrust fault development and demonstrated the deformation process of strata with the creation of high-angle branching faults from low-angle thrust. Thus, the combination of low-angle decollement plane with high-angle branching faults is a universal phenomenon, and it is important to assess the earthquake and tsunami potential of the high-angle branching faults around a subduction zone in addition to that of plate-boundary-type earthquakes.

According to the Nakata map, the height of the tectonic scarp along the high-angle branching faults varies from 2 to 3 km along the strike. We considered that the height difference of the tectonic scarp corresponds to the difference in cumulative displacement of previous repeating earthquakes and therefore is related to both the coseismic displacement and relative activity of the faults. From a viewpoint of interpretation of tectonic landforms, it is characterized that type Y section shows the highest tectonic scarp in the Nakata map, and types A to D sections show the difference of tectonic scarp height in descending order.

4.3   Fault Parameters

The parameters of location, length, strike, dip angle, depth of fault upper edge, and fault width for the earthquake source fault model were determined using the fault distribution and character of the tectonic scarps in the Nakata map. First, the location, length, and strike were determined from the type differences in the Nakata map from ten linear sections with different lengths between 30 and 80 km and the individual strike (Fig. 4.2a) Second, the dip angle and the depth of the fault upper edge were examined by comparing the Nakata map with the seismic profile records (e.g., Kato [3]; Fig. 4.1c). The dip angle of type Y section in the Nakata map corresponds to the decollement of the plate boundary, and the average of 14°W was assigned to the parameters from several seismic profile records ranging from 12°W to 16°W. The depth of the fault upper edge was assigned as 0 km since the decollement of the plate boundary reaches the sea bottom.

The dip angles of types A to D sections of high-angle branching faults were assigned as 26°W from several seismic profile records ranging from 25°W to 28°W. The tectonic landforms of these sections display a humped form with a relatively long wavelength of 20 km, and such hump-shaped tectonic landforms have been observed empirically when the upper fault edge was buried beneath the subsurface. Thus, the depth of the fault upper edges of these types was 5 km beneath the tectonic scarps on the sea bottom, as shown in Fig. 4.1c. For the fault width, the parameters of low-angle decollement plane (e.g., Fujii and Satake [5], Fig. 4.2a) for the 2011 Tohoku-oki earthquake were referred since the high-angle branching faults in this study converged to the decollement of the plate boundary (Fig. 4.1c).

Subsequently, the average slip value of 11 m was calculated from the empirical equation of fault area and average displacement given in "A Formula for the Prediction of Strong Ground Motion of Subduction-Type Earthquakes" [6]. Finally, the slip distribution of 25 sub-faults was assigned using a trial-and-error method under the following four constraint conditions: (1) the total seismic moment is preserved, (2) the maximum slip is almost twice the average slip according to the formula of HERP [6], (3) the area of maximum slip is almost 20 % of the total fault area according to the formula of HERP [6], and (4) the slip distribution of the section of decollement is close to that of the 2011 Tohoku-oki earthquake since our model shared the deep part of the low-angle fault plane with that of the 2011 Tohoku-oki earthquake as the decollement of the plate boundary.

The resulting model parameters (Model1) are summarized in Table 4.1 and Fig. 4.2b. The amount of slip and the area of the asperity of the formula [6] were

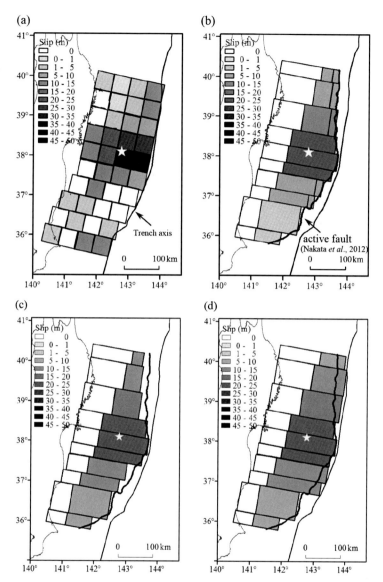

Fig. 4.2 Earthquake source fault models. (**a**) The 2011 Tohoku-oki earthquake from Fujii and Satake [5]; (**b**) Model1, the high-angle branching fault model of this study; (**c**) Model2, sub-faults of Model1 without high-angle branching part; (**d**) Model3, sub-faults of Model2 and extended sub-faults to the sea bottom with same low-angle and same slip amount. The epicenter of the 2011 Tohoku-oki earthquake is indicated by an *asterisk*

in close agreement with those of the shallow part of the sub-faults along the type Y section in the Nakata map. It is important to note that the location and depth of the upper fault edge of the type A to D sections as high-angle branching faults in

Table 4.1 Fault parameters used in the earthquake source fault model (Fig. 4.2b–d)

ID	Latitude (deg)	Longitude (deg)	Depth (km)	Strike (deg)	Dip (deg)	Rake (deg)	Length (km)	Width (km)	Slip (m)
1	40.1	144.0	5	190	27	90	36	20	6
2	40.2	143.7	15	190	14	90	36	32	6
3	40.2	143.4	23	190	14	90	36	121	0
4	39.8	143.9	5	190	27	90	71	20	10
5	39.8	143.7	15	190	14	90	71	49	10
6	39.9	143.1	27	190	14	90	71	109	0
7	39.1	143.9	5	190	27	90	73	20	14
8	39.2	143.6	15	190	14	90	73	78	14
9	39.3	142.8	34	190	14	90	73	92	0
10	38.5	143.8	0	190	14	90	46	129	21
11	38.7	142.3	31	190	14	90	46	74	0
12	38.0	143.8	0	190	14	90	46	145	21
13	38.3	142.2	35	190	14	90	46	73	0
14	37.6	143.6	0	190	14	90	30	140	21
15	37.9	142.1	34	190	14	90	30	70	0
16	37.4	143.3	5	190	27	90	45	20	10
17	37.4	143.1	15	190	14	90	45	111	10
18	37.6	141.9	42	190	14	90	45	61	0
19	37.0	143.0	5	190	27	90	35	20	6
20	37.0	142.8	15	190	14	90	35	111	6
21	37.2	141.6	42	190	14	90	35	54	0
22	36.7	142.6	0	190	14	90	78	113	5
23	36.9	141.3	27	190	14	90	78	55	0
24	36.1	142.0	0	190	14	90	34	74	5
25	36.2	141.2	18	190	14	90	34	56	0
a	40.10	144.21	0.10	190	14.0	90	36	75	4
b	39.75	144.22	0.10	190	14.0	90	71	96	7
c	39.11	144.05	0.10	190	14.0	90	73	117	12
d	37.34	143.54	0.10	190	14.0	90	45	155	8
e	36.95	143.30	0.10	190	14.0	90	35	156	5

Fig. 4.2b differed from those of the 2011 Tohoku-oki earthquake (e.g., Fujii and Satake [5], Fig. 4.2a). In addition, the slip amount of type Y section and type A to D sections are empirically average values compared to the exceptionally large value estimated for the 2011 Tohoku-oki earthquake (Fig. 4.2a) compared to the empirical average value of the type Y section in Fig. 4.2b. In order to clarify the influence of high-angle branching faults in this study, two additional models are compared; one is Model2 displayed in Fig. 4.2c consisting of Model1 sub-faults without high-angle branching parts (ID; 1, 4, 7, 16, 19 in Table 4.1) and the other is Model3 displayed in Fig. 4.2d replacing high-angle branching parts with extended low-angle sub-faults to the sea bottom with same seismic moment (ID; a–e in

Table 4.1). The moment magnitude and average slip amount of these three models are calculated as Mw8.9 and 6.6 m in Model1, Mw8.8 and 6.0 m in Model2, and Mw8.9 and 5.5 m in Model3.

4.4 Tsunami Simulation and Results

A simulation of tsunami propagation was performed using the TSUNAMI-K software developed by Kozo Keikaku Engineering, Inc., which applies an equation of motion and an equation of continuity including nonlinear long-wave theory with an advective term and a sea-bottom friction term in the Staggered Leapfrog Method. The initial condition of tsunami production was derived from the static tectonic movement calculated from the fault displacement model using the parameters in Table 4.1. The Manning roughness coefficient was assigned the typical value of 0.025 $[m^{(-1/3)}$ s]. A Digital Elevation Model of 50-m grid interval for onshore areas from the Geospatial Information Authority of Japan and a Digital Bathymetry Model of 500-m grid interval for offshore areas from the Hydrographic and Oceanographic Department, Japan Coast Guard were used in the simulation. The grid size for the simulation was assigned as 1350 m, and the duration period for the tsunami simulation was 2 h with an interval of 1 s to satisfy the stability condition of the simulation.

Several indices, such as tsunami height, trace height, and run-up height, were adopted from observed tsunami evaluations. However, observational data from onshore and shallow-sea points are directly influenced by local undulation and landforms and were inappropriate for comparison with the simulation results of the Digital Bathymetry Model with its 500-m resolution. Therefore, we compared our simulation results with observed tsunami records from sea-bottom pressure sensors located at points P02 and P06 [7] and tsunami recorder at points TM1 and TM2 [8] in Fig. 4.1a situated off the coast of Iwate to demonstrate the maximum tsunami height. Figure 4.3 shows the comparison of tsunami heights of Model1, Model2, and Model3. The maximum heights obtained from high-angle branching fault simulation (black line) were 2.8 m at P02, 2.9 m at P06, 2.7 m at TM1, and 3.0 m at TM2. These values are approximately 60 % of the observed maximum height of 5 m recorded at the 2011 Tokoku-oki earthquake. However, when we used the estimated fault parameters of Fig. 4.2a in our tsunami simulation software to simulate the 2011 Tokoku-oki earthquake, the maximum heights are 0.8 m smaller with Model1 at TM1 and TM2 though almost same with Model1 at P02 and P06. This means that the sea-bottom deformations of Model1 with high-angle branching fault and average slip amount calculated using the fault dislocation model show almost equivalent potential as the estimated model for the 2011 Tokoku-oki earthquake if we consider the accuracy and limitations of the tsunami simulations. The comparison between three models and observed tsunami height at TM1 and TM2 where these recorders are located at the back of the sections of high-angle branching faults also suggests that earthquakes produced by branching faults with

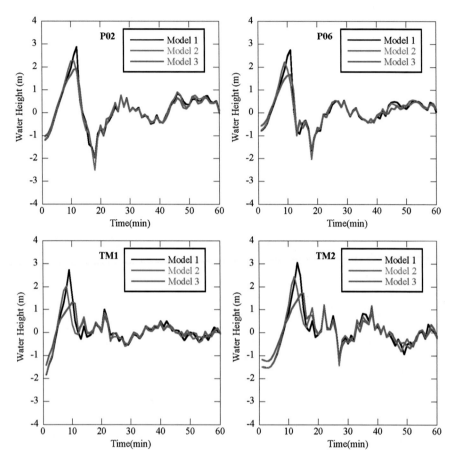

Fig. 4.3 Comparison of simulated tsunami height at points P02, P06, TM1, and TM2 (Fig. 4.1) where the observed data of the 2011 Tohoku-oki earthquake are recorded. Model1, *black line*; Model2, *orange line*; Model3, *green line*

average slip possess a tsunami potential comparable to that produced by the exceptionally large slip of the 2011 Tohoku-oki earthquake at several points along the coast. In other words, the observed severe tsunami of the 2011 Tohoku-oki earthquake may not be exceptional for the eastern coast of Tohoku District. The situation is similar along the eastern coast of Japan where faults exist on the continental slope parallel to the trench axis, and the tsunami potential of the high-angle branching faults should be considered in the design of disaster prevention measures.

4.5 Summary and Conclusion

The potential of a tsunami produced by a future earthquake due to active faults on the continental slope was identified from the 3D tectonic landform analysis compiled in the Nakata map. Some of these faults display relatively high-angle dipping estimated from seismic profiling records. Thus, the simulated tsunami potential is almost equal to that of the 2011 Tohoku-oki earthquake under limitations of the simulations since faults of high angle with average slip produce effects similar to sea-bottom deformation compared to low-angle faults with exceptionally large slip as in the 2011 Tohoku-oki earthquake.

These active faults on the continental slope are recognized not only along the Japan Trench examined in this study but also the Nankai Trough in the western part of Japan. To evaluate the tsunami potential including both "plate-boundary megathrust" and high-angle branching fault type earthquakes, the Digital Bathymetry Model should be developed further, and the use of combined comprehensive interpretations of geophysical and geomorphological analysis should be promoted.

Acknowledgment The authors thank Prof. Nakata of Hiroshima University for provision of data. This research was supported by Grant-in-Aid for Scientific Research (1940300199).

References

1. Nakata T et al (2012) Active faults along Japan Trench and source faults of large earthquakes. http://www.jaee.gr.jp/event/seminar2012/eqsympo/pdf/papers/34.pdf. 19 Dec 2012
2. Okamura Y (2012) Examine the 2011 Tohoku-Oki earthquake from geology (in Japanese). J Earthq 54:1–12
3. Kato Y (2000) Mega-submarine slide and geological structure off Fukusima. Abstracts Japan Earth and Planetary Science Joint Meeting: Sj-P004
4. Ueta K, Tani K (1999) Deformation of Quaternary deposits and ground surface caused by bedrock fault movements (Part 2)-normal and reverse fault model tests- (in Japanese). Rep Cent Res Inst Electr Power Ind U98048:1–40
5. Fujii Y, Satake K (2011) Tsunami source models of the 2011/03/11 Tohoku-oki earthquake (preliminary result). http://iisee.kenken.go.jp/staff/fujii/OffTohokuPacific2011/tsunami_ja.html
6. HERP (2008) Recipe for predicting strong ground motion from subduction earthquake (in Japanese). http://www.jishin.go.jp/main/kyoshindo/05dec_miyagi/recipe.pdf
7. Saito T et al (2011) Tsunami source of the 2011 Tohoku-oki earthquake, Japan: inversion analysis based on dispersive tsunami simulations. Geophys Res Lett 38. doi:10.1029/2011GL049089
8. Maeda T et al (2011) Significant tsunami observed at the ocean-bottom pressure gauges at 2011 Off the Pacific Coast of Tohoku Earthquake. Earth Planets Space. doi:10.5047/eps.2011.06.005

Chapter 5
Extensive Area of Topographic Anaglyphs Covering Inland and Seafloor Derived Using a Detailed Digital Elevation Model for Identifying Broad Tectonic Deformations

Hideaki Goto

Abstract Topographic anaglyph images were viewed with red-cyan glasses enabled to recognize topographic relief features easily. Anaglyphs produced from digital elevation model (DEM) data are a very effective technique to identify tectonic geomorphology. The aim of this paper was to introduce an extensive area of topographic anaglyph images produced from the 5-m-mesh and 10-m-mesh inland DEM of Geospatial Information Authority of Japan, as well as the 1-s-mesh DEM on the seafloor. In this paper, we present two examples which show that the extensive area of anaglyph produced from combined detailed DEM is advantageous for identifying broad tectonic geomorphology near a coastal area, as well as in urban areas, to view "naked" topography exaggerated vertically. For instance, the NW-SE trending active flexure scarp on the Musashino surface to the north of Tokyo Metropolis has been identified by means of interpretation of these images. The tectonic deformation on the shallow seafloor near Kisakata has also been identified, where the emergence of the lagoon associated with the Kisakata earthquake (M7.0) of 1804 was recorded in the historical documents. When anaglyphs from detailed DEM are extensive and have emphasized vertical exaggeration, they are valuable for recognizing long-wave (one kilometer to several hundred meter scaled) deformations.

Keywords Active fault • Anaglyph • Tectonic geomorphology • Digital elevation model • Submarine topography • Kisakata • Musashino surface • Arakawa fault

5.1 Introduction

Topographic anaglyph images viewed with red-cyan glasses enables one to recognize the features of topographic relief in three dimensions. Anaglyphs produced from a digital elevation model (DEM) are very useful material for identifying

H. Goto (✉)
Hiroshima University, 1-2-3, Kagamiyama, Higashihiroshima, Hiroshima 739-8522, Japan
e-mail: hgoto@hiroshima-u.ac.jp

© The Author(s) 2016
K. Kamae (ed.), *Earthquakes, Tsunamis and Nuclear Risks*,
DOI 10.1007/978-4-431-55822-4_5

tectonic geomorphology. For instance, Goto and Sugito [1] revealed the discovery of several unknown inland active faults. Furthermore, Izumi et al. [2] revealed the distribution of submarine active faults along the eastern margin of the Japan Sea.

The Geospatial Information Authority of Japan (GSI) has published inland DEM data since the Basic Act on Promotion of Utilization of Geographical Information implemented in 2007. Goto [3] introduced anaglyph images to study inland geomorphology of Japan produced by the use of the 10-m-mesh DEM of GSI. However, high-resolution topographic anaglyph images covering both onshore and offshore areas have not been published because of a lack of detailed topographic data. Goto [4] has presented the topographic anaglyph images of the seafloor around Japanese Islands that were produced by using the 1-s-mesh (approximately 30 m) DEM provided by the Japan Coast Guard.

The aim of this paper is to introduce extensive area of topographic anaglyph images produced from the 5-m-mesh and 10-m-mesh inland DEM of GSI, as well as the 1-s-mesh DEM (Goto [4]) on the seafloor. We identified active faults that have deformed alluvial plains, terrace surfaces, and seafloors by means of interpretation of these anaglyph images. Smaller fault scarps and fault-related broad deformations were newly identified in numerous inland sites as well as on the shallow seafloor. In this paper, we present two examples that show that the extensive area of topographic anaglyphs produced from combined detailed DEM is quite advantageous for identifying broad tectonic deformation near a coastal area, as well as in urban areas, when the "naked" topography is exaggerated vertically.

5.2   Data and Methods

Detailed DEM of the shallow seafloor obtained from direct data acquisition, such as acoustic prospecting/seismic profiling, is only available for a limited area of the coast. Thus, we reprocessed the 1-s-mesh DEM from digital bathymetric charts (M7000 series), with 1- to 2-m interval counters of the Japan Hydrographic Association (Goto [4]) (Fig. 5.1 A–D). Currently, the published area for inlands at 5-m-mesh DEM of GSI is still about 45 % of the territory of Japan, as of November 2013. Therefore, the 5-m- and 10-m-resolution meshes that are at present the most detailed DEM open to the public were combined for inland geomorphology (Fig. 5.1 E, F).

We imported both inland and seafloor data that ranges from 2° in longitude and 1° 20′ in latitude to a Simple DEM Viewer®. Then, we produced an anaglyph image to be overlapped on the base map of the black and white slope shading map and the black and white shaded relief map. We produced 65 map sheets under these conditions with less than 35,000 pixels on one side of each sheet. We also provided topographic anaglyph overlaid active fault maps (Nakata and Imaizumi eds. [5]) to facilitate a reexamination of active fault geomorphology (Fig. 5.2).

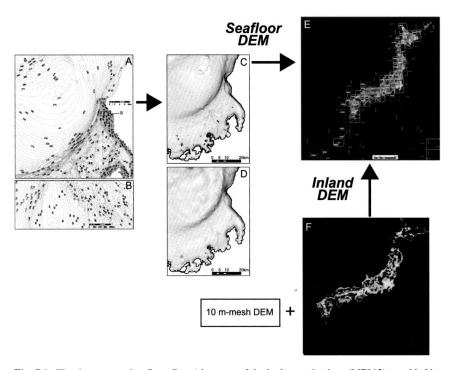

Fig. 5.1 The data processing flow. *Box A* is a part of the bathymetric chart (M7012) provided by the Japan Coast Guard from which we reprocessed 1-s-mesh DEM. The interval of counters on the chart is different from place to place. The distance between counters is under 50 m as shown in the *Box B*. Anaglyph (*C*) produced from 1-s-mesh (ca.30 m) DEM reprocessed from bathymetric chart (M7012) is quite finer than anaglyph (*D*) produced from 500-m-mesh DEM (J-EGG500). The published area of 5-m-mesh DEM of GSI is limited as of November 2013 (*F*). The 65 sheets of detailed topographic anaglyph within the range of 2° in longitude and 1°20′ in latitude are produced (*E*)

5.3 Fault-Related Broad Deformations in an Urban Area Identified on the Anaglyph

Airborne LiDAR (light detection and ranging) surveying has been developed to detect detailed geomorphic features (e.g., Nelson et al. [6]). Airborne LiDAR has the capability of revealing the "bare earth," with vegetation and buildings removed, because the ground emits a laser pulse that can be separated from canopy returns through a filtering process. Thus, these data can be used to explore quantitatively the characteristics of tectonic geomorphology (e.g., Arrowsmith and Zielke [7]). The data also revealed small-scale fault scarps in the urban area (e.g., Kondo et al. [8]) as well as on the mountain slopes beneath the dense forest vegetation (e.g., Lin et al. [9]). However, fault-related broad deformations were invisible on the usual topographical maps (slope shade map, hill shade map, and contour map), even if airborne LiDAR data was adopted. Topographical anaglyphs produced from

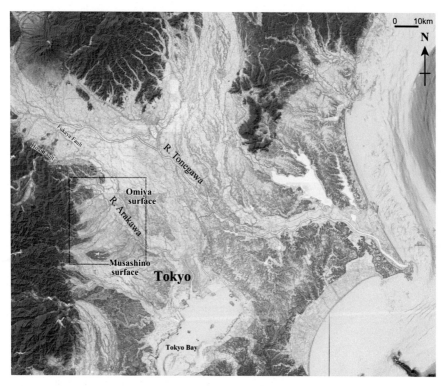

Fig. 5.2 Extensive area of a topographic anaglyph around Tokyo Metropolis. It is one of the 65 sheet maps produced from 5-m-mesh and 10-m-mesh DEM of inland of GSI and 1-s-mesh (ca.30 m) DEM of the seafloor. The *dotted-line box* indicates the area of Fig. 5.3

DEM based on LiDAR data are easily observed in three dimensions like stereo-paired aerial photographs, when wearing red-cyan glasses. Broad deformation related to recent faulting could be identified on the extensive area of the topographic anaglyph because it can be emphasized in vertical exaggeration and rescaled on a PC monitor. Here, newly identified tectonic geomorphology to the north of Tokyo Metropolis is shown as an example (Fig. 5.2).

The Arakawa River flows from the Kanto Mountains into Tokyo Bay through the northwestern part of the Kanto basin, where the Pleistocene terraces are well developed. The terraces situated on the left and right banks in the middle reaches of the river are called the Omiya surface and the Musashino surface, respectively (Fig. 5.2). Kaizuka [10] suggested that the northern part of the Musashino surfaces was inclined to the northeast, compared to the east and south slope of this surface, based on an analysis of a contour line map. For this reason, the NW-SE trending east side up the Arakawa fault that is buried beneath the alluvial plain would be extended along the southwestern side of the Omiya surface (Kaizuka [10]). On the contrary, Hirouchi [11] supposed that the displacement of surfaces across the Arakawa fault was not observed, based on the topographic profiles on the

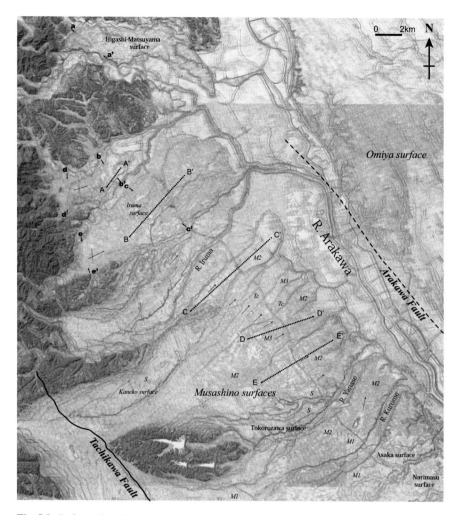

Fig. 5.3 Deformation of the Musashino surfaces to the north of Tokyo Metropolis showing on the detailed topographic anaglyph. The *purple arrows* indicate the flexure scarps on the Musashino surface (Musashino flexure zone). The newly identified small scarps related to recent faulting are marked with small arrows (*a–a'* to *e–e'*). The small scarps (*b–b'* to *c–c'*) on this figure are called the Tsurugashima fault. The active fault lines depicted in the previous paper show *black lines*

Musashino surface and Omiya surface. The Headquarters for Earthquake Research Promotion [12] assessed that the east-side-up buried fault was not distributed along the Arakawa River. Therefore, the geomorphological evidence of tectonic crustal movement has not been clarified in this area.

When we interpreted the detailed topographic anaglyphs of this area, the west-side-up several-hundred-meter-long convex slopes were found on the northern part of the Musashino surfaces (Fig. 5.3). These scarps must have been flexure scarps related to recent faulting, because these NW-SE trending scarps were detected at

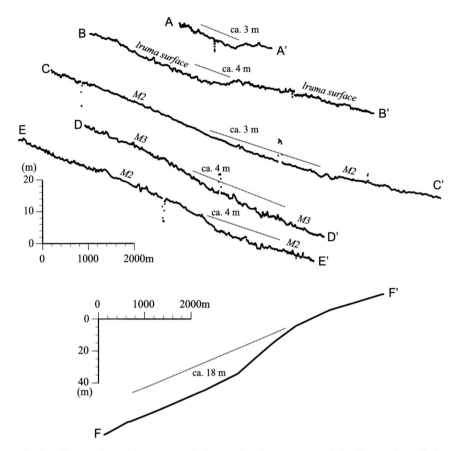

Fig. 5.4 Topographic profiles across the Musashino flexure zone and the Tsurugashima fault, based on the DEM. The measured lines are shown in the Figs. 5.3 and 5.5

different heights of Musashino surfaces (M2, M3) and Tachikawa surface (Tc) (Figs. 5.3 and 5.4) and are designated here as the Musashino flexure zone.

To the northwest of this zone, the small fault scarps facing the mountain (the Tsurugashima fault) were also newly identified on the Iruma surface (Fig. 5.3 b–b', c–c'). Other small fault scarps on the surfaces are also detected using the anaglyphs (Fig. 5.3 a–a', d–d', e–e'). The Musashino flexure zone extends as a series of the Tsurugashima fault although showing the reverse sense of the downthrown side. The upthrown side along the strike-slip fault is, in general, located on the relative strike-slip motion side of the faulted block (e.g., Nakata and Goto [13]). Thus, the active tectonic movement forming these features may have a left-lateral component, although at present we failed to find any geomorphological evidence for strike-slip faulting along these scarps. Furthermore, a series of the Musashino flexure zone and the Tsurugashima fault might relate to the Hirai left-lateral fault system situated to the northwest along the foot of Kanto Mountain (Fig. 5.2).

5.4 Coastal Geomorphology on the Anaglyph

Coastal geomorphology, such as the distribution of the Holocene marine terraces and differences in the heights of the shoreline of the Pleistocene marine terraces along the coastline, could be a clue for not only sea-level changes but also active tectonics (Earth crustal movements). However, the number of published geomorphologic reports is limited because of the lack of extensive topographic maps covering both the onshore and offshore area seamlessly. One of the technical methods to overcome such difficulties is to make seamless anaglyph images as shown in this paper. Such an approach to tectonic topography including for unknown active faults come to light around the coastline. Here, submarine active tectonic geomorphology off Kisakata, Akita Prefecture, is shown as an example, in order to indicate the usefulness of the anaglyph images for better understanding of active tectonics in a coastal region.

Tectonic deformation on the shallow seafloor has been newly identified near Kisakata, where the emergence of the lagoon associated with the Kisakata earthquake (M7.0) of 1804 was recorded in the historical documents (Usami [14]). According to the historical pictures and documents, the hummocky hills caused by the debris avalanche from Mt. Chokaisan were distributed in the lagoon around Kisakata before the earthquake. We could observe so many different sizes of hills to the northwestward of Mt. Chokaisan on the anaglyph (Fig. 5.5). We could also recognize the paleo-lagoon in Kisakata by the scattered small hills surrounding the flat plains.

To the 10–14 km west of Kisakata and Yuri-Honjo, the NS trending continental shelf slope divides the Tobishima basin from the continental shelf. Although several anticlinal axes of deformed Quaternary sediment are estimated to be striking parallel to this slope beneath the continental shelf (Osawa and Suda [15]), no surface evidence of tectonic deformation has not been recognized off Kisakata.

However, the NNE trending steep flexure slope on the continental shelf along the coastline neighboring to the paleo-lagoon is illustrated in the seamless anaglyph image (Fig. 5.5). The slope extends for at least 15 km in the middle of the 10-km-wide shallow seafloor, with a gentle sloping toward WNW. These topographic features are consistent with subsurface structure recorded by the acoustic exploration (Geological Survey of Japan [16]). The active anticline axis parallel to the coastline was identified in the south extension of this slope by seismic reflection profiling (Horikawa et al. [17]). This long-wave deformation would most likely be caused by a subsurface fault, which is the most probable candidate for the source of the 1804 earthquake, because the uplifted area associated with the earthquake (Usami [14]) is parallel to this deformation.

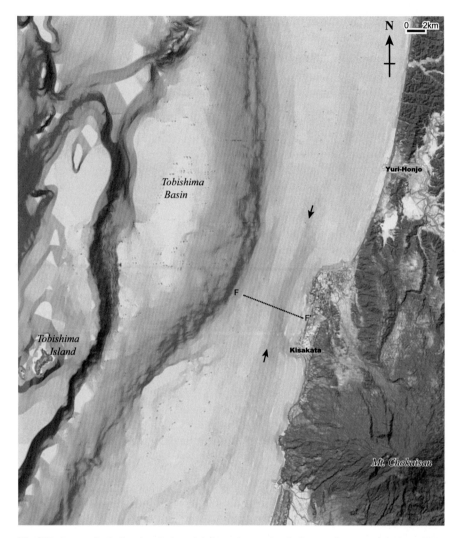

Fig. 5.5 An anaglyph showing the broad deformation on the shallow seafloor near Kisakata, Akita Prefecture. The newly identified broad scarps related to recent faulting are marked with *small arrows*

5.5 Conclusions

We produced detailed topographic anaglyphs from inland DEM measured by LiDAR data as well as 1-s-mesh DEM from digital bathymetric charts. We could easily observe macro- to micromorphology in three dimensions on a seamless anaglyph image covering both onshore and offshore using red-cyan glasses. Although it is difficult to detect a broad deformation related to recent faulting on any conventional maps, we could easily recognize such a deformation on anaglyphs when it is emphasized in vertical exaggeration. Future research combining seamless anaglyph images and well-trained personnel, to identify tectonic geomorphology around nuclear power plant sites, will provide important datasets for seismic safety evaluation.

5.6   Acknowledgment

My heartfelt appreciation goes to Dr. Takashi Nakata, Professor Emeritus at Hiroshima University whose comments and suggestions were of inestimable value for my study. I am also indebted to Yoshiaki Katayanagi, who kindly modified his software (Simple DEM viewer) for our analysis. I have also had the support and encouragement of Prof. Mitsuhisa Watanabe, Prof. Takashi Kumamoto, Prof. Yasuhiro Kumahara, Prof. Nobuhisa Matsuta, Dr. Nobuhiko Sugito, and Prof. Takahiro Miyauchi. This work was funded by JSPS KAKENHI grant numbers 25350428 and 23240121.

References

1. Goto H, Sugito N (2012) Fault geomorphology interpreted using stereoscopic images produced from digital elevation models. E-journalGEO 7:197–213. http://dx.doi.org/10.4157/ejgeo.7. 197 (in Japanese with English abstract)
2. Izumi N, Nishizawa A, Hirouchi D, Kido Y, Nakata T, Goto H, Watanabe M, Suzuki Y (2014) 3D bathymetric image of the eastern margin of the Sea of Japan based on 3-second grid DEM. Rep Hydrogr Oceanogr Res 51:127–139. http://www1.kaiho.mlit.go.jp/GIJUTSUKOKUSAI/ KENKYU/report/rhr51/rhr51-TR10.pdf (in Japanese with English abstract)
3. Goto H (2012) Anaglyph images of Japan produced from digital elevation model: explanatory text and sheet maps. Hiroshima University Studies Graduate School of Letters, Special Issue 72: 69p. http://ir.lib.hiroshima-u.ac.jp/00034345 (in Japanese with English abstract)
4. Goto H (2013) Submarine anaglyph images around Japan Islands based on bathymetric charts: explanatory text and sheet maps. Hiroshima University Studies Graduate School of Letters, Special Issue 72: 69p. http://ir.lib.hiroshima-u.ac.jp/00035603 (in Japanese with English abstract)
5. Nakata T, Imaizumi T (eds) (2002) Digital active fault map of Japan. Tokyo Press, Tokyo, 68 p (in Japanese with English abstract)
6. Nelson AR, Johson SY, Kelsey HM, Wells RE, Sherrod BL, Pezzopane SK, Bradley L, Koehler RD, Buchnam RC (2003) Later Holocene earthquakes on the Toe Jam Hill Fault, Seattle fault zone, Bainbridge Island, Washington. Geol Soc Am Bull 115:1388–1403
7. Arrowsmith JR, Zielke O (2009) Tectonic geomorphology of the San Andreas Fault zone from high resolution topography: an example from the Cholame segment. Geomorphology 113: 70–81

8. Kondo H, Toda S, Okumura K, Takada K, Chiba T (2008) A fault scarp in an urban area identified by LiDAR survey: a case study on the Itoigawa–Shizuoka Tectonic Line, central Japan. Geomorphology 101:731–739

9. Lin Z, Kaneda H, Mukoyama S, Asada N, Chiba T (2013) Detection of subtle tectonic geomorphic features in densely forested mountains by very high resolution airborne LiDAR survey. Geomorphology 182:104–115. doi:10.1016/j.geomorph.2012.11.001

10. Kaizuka S (1957) Deformation of the diluvial upland Musashino, and its significance on the movements of Kanto tectonic basin. Quat Res 1:22–30 (in Japanese)

11. Hirouchi D (1999) A Reconsideration of the activity of the Arakawa fault deduced from the longitudinal profiles of Late Quaternary terraces in the Musashino and Omiya uplands. Geogr Rev Jpn 72:335–344 (in Japanese with English abstract)

12. The Headquarters for Earthquake Research Promotion (2004) Long-term evaluation of the Arakawa fault. http://www.jishin.go.jp/main/chousa/04aug_arakawa/index.htm (in Japanese)

13. Nakata T, Goto H (1998) New geometric criteria for active fault segment; Fault branching and dip-slip distribution pattern along strike-slip faults. Act Fault Res 17:43–53. http://ir.lib. hiroshima-u.ac.jp/00025450 (in Japanese with English abstract)

14. Usami T (2008) Materials for comprehensive list of destructive earthquakes in Japan, Revised and enlarged edn. Tokyo University Press, Tokyo (in Japanese). http://www.utp.or.jp

15. Osawa A, Suda Y (1980) Geological map of Japan 1:200,000. Geological Survey of Japan, Akita/Oga

16. Geological Survey of Japan (2015) Database of offshore geologic structure. https://gbank.gsj. jp/marineseisdb/index_E.html

17. Horikawa H, Okamura Y, Murakami F (2011) Seismic reflection profiling of the shallow structure of the Sakata-oki Uplift, Northeast Japan. Ann Rep Act Fault Paleoearthq Res 11: 83–96 (in Japanese with English abstract)

Part II
Seismic Source Modeling and Seismic Motion

Chapter 6
Relation Between Stress Drops and Depths of Strong Motion Generation Areas Based on Previous Broadband Source Models for Crustal Earthquakes in Japan

Toshimi Satoh and Atsushi Okazaki

Abstract To aim at the advancement of strong motion predictions, we develop empirical relations between stress drops on strong motion generation areas (SMGAs) and depths of SMGAs based on previous broadband source models estimated by the empirical Green's function method. A total of 25 source models for 13 crustal earthquakes of Mw from 5.7 to 6.9 in Japan are used in this study. It is found that stress drops on SMGAs for reverse faults are larger than those for strike-slip faults on average. The average stress drops are 21.2 MPa, 13.3 MPa, and 18.0 MPa for reverse, strike-slip, and all types of faults, respectively. In the derived empirical relation for all types of faults, the stress drops increase by about 1 MPa every 1 km in depth. This depth dependency is similar to the relation between stress drops on asperities and the depths of asperities derived by Asano and Iwata (Pure Appl Geophys, 168:105–116, 2011), and the absolute value is 4 MPa larger than that by Asano and Iwata (Pure Appl Geophys, 168:105–116, 2011). The depth dependency of stress drops for reverse faults is stronger than that for strike-slip faults. The total area of SMGAs is about 0.8 times of the total area of asperities by Somerville et al. (Seismol Res Lett, 70:59–80, 1999). The result can be interpreted by frequency-dependent source radiations, since asperities are estimated from longer-period (>2 s) strong motions than SMGAs, which are mainly estimated from strong motions in the period range from 0.1 to 5 s.

Keywords Stress drop • Strong motion generation area • Crustal earthquake • Focal depth • Empirical Green's function method • Reverse fault • Strike-slip fault • Source model

T. Satoh (✉)
Shimizu Corporation, 4-17 Etchujima 3-chome, Koto-ku, Tokyo 135-8530, Japan
e-mail: toshimi.satoh@shimz.co.jp

A. Okazaki
Kansai Electric Power Company, 6-16, Nakanoshima 3-chome, Kita-ku, Osaka-shi, Osaka 530-0005, Japan

© The Author(s) 2016
K. Kamae (ed.), *Earthquakes, Tsunamis and Nuclear Risks*,
DOI 10.1007/978-4-431-55822-4_6

6.1 Introduction

Kamae and Irikura [1] estimated a broadband source model for the 1995 Hyogoken-Nanbu earthquake (Mw6.9) to fit near-field strong motion records by forward modeling using the empirical Green's function method (e.g., Irikura [2]). Since Kamae and Irikura's pioneering work, many researchers have estimated broadband source models in the period range from about 0.1 to 5 s for the other big earthquakes using the empirical Green's function method. The broadband source models were composed of a few strong motion generation areas (SMGAs).

Irikura and Miyake [3] proposed the recipe for strong motion predictions. In the recipe, the source model was expressed by a few rectangular asperities and surrounding background areas. Here asperities were characterized from long-period heterogeneous kinematic slip models estimated by waveform inversion method using strong motion records in the period longer than about 2 s [4]. Asano and Iwata [5] studied on the relations between stress drops on asperities and depths of asperities for crustal earthquakes. Miyake et al. [6] showed that asperities coincide to SMGAs defined as areas that mainly generate strong ground motions. However, the period ranges to estimate asperities based on the waveform inversion results are relatively longer than those to estimate SMGAs by the empirical Green's function method. Therefore, we develop empirical relations between stress drops on SMGAs and depths of SMGAs based on the previous broadband source models estimated by the empirical Green's function method for crustal earthquakes in Japan to aim at the advancement of strong motion predictions.

6.2 Data

Data used in this study are shown in Table 6.1 [1, 6–34] and Fig. 6.1. A total of 22 articles on SMGAs [1, 6, 13, 16–34] for 13 crustal earthquakes of the moment magnitude Mw from 5.7 to 6.9 in 1995 to 2011 are used. The numbers of the strike-slip, reverse, and normal faults are six, six, and one, respectively. We independently treat each source model for the same earthquake, and so the total 25 source models are examined. We also independently treat each strong motion generation area. The stress drops on SMGAs estimated by Miyake et al. [6] are calculated assuming the single-asperity model. The others are calculated assuming single-crack models.

6.3 Results

Figure 6.2 shows the relations between stress drops on SMGAs and the top, center, and bottom depths of SMGAs for strike-slip faults and reverse faults. The number of SMGAs of each earthquake is one to three except for five by Hirai et al. [16]. The

Table 6.1 List of earthquakes and references [1, 6–34]

No.	Name of earthquake	Mw	References on M_o and Mw	Fault type	References on strong motion generation areas
1	1995 Hyogoken-Nanbu	6.9	Sekiguchi et al. [7]	Strike slip	Kamae and Irikura [1]
					Hirai et al. [16]
2	1997 Kagoshima-ken-Hokuseibu (March)	6.0	Kuge et al. [8]	Strike slip	Miyake et al. [6]
3	1997 Yamaguchi-ken Hokubu	5.8	F-net	Strike slip	Miyake et al. [6]
4	1998 Iwate-ken Nairiku Hokubu	5.9	F-net	Reverse	Miyake et al. [6]
5	2000 Tottori-ken Seibu	6.7	Iwata and Sekiguchi [9]	Strike slip	Ikeda et al. [17]
					Satoh and Irikura [18]
6	2004 Niigata-ken Chuetsu	6.6	Hikima and Koketsu [10]	Reverse	Kamae et al. [19]
					Satoh et al. [20]
7	2004 Rumoi	5.7	F-net	Reverse	Maeda and Sasatani [21]
8	2005 Fukuoka-ken Seiho-oki	6.6	Asano and Iwata [5]	Strike slip	Suzuki and Iwata [22]
					Miyake et al. [23]
					Satoh and Kawase [24]
					Ikeda et al. [25]
9	2007 Noto Hanto	6.6	Horikawa [12]	Reverse	Kurahashi et al. [26]
					Maeda et al. [27]
					Ikeda et al. [25]
10	2007 Niigata-ken Chuetsu-oki	6.6	Shiba [13]	Reverse	Irikura [28]
					Shiba [13]
					Yamamoto and Takenaka [29]
					Kawabe and Kamae [30]
11	2008 Iwate-Miyagi Nairiku	6.8	Horikawa [14]	Reverse	Kamae [31]
					Irikura and Kurahashi [32]
12	2011 Shizuoka-ken Tobu	5.9	F-net	Strike slip	Somei et al. [33]
13	2011 Fukushima-ken Hamadori	6.6	Hikima [15]	Normal	Somei et al. [34]

stress drops are less than 30 MPa except for those for the 2007 Noto Hanto earthquake estimated by Maeda et al. [27]. Although source models for the same earthquake are different among researchers, large SMGAs tend to be located at deep places.

Figure 6.3a shows the relations between the stress drops and the center depths of SMGAs for all types of earthquakes including a normal-fault earthquake and regression relations. The average stress drops are 21.2 MPa with standard error of 9.2 MPa, 13.3 MPa with standard error of 5.3 MPa, and 18.0 MPa with standard error of 8.6 MPa for reverse, strike-slip, and all types of faults, respectively. The average stress drop on SMGAs of 18.0 MPa in this study is larger than average

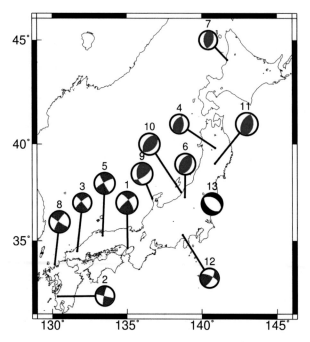

Fig. 6.1 Epicenters by JMA and focal mechanisms by the Global CMT Project for No.1 earthquake and by F-net for the other earthquakes

○ 1995 Hyogo-ken Nanbu(Kamae and Irikura,1998)
+ 1995 Hogo-ken Nanbu(Hirai et al.,2006)
■ 1997 Yamaguchi-ken Hokubu(Miyake et al.,2003)
■ 1997 Kagoshima-ken Hokuseibu(Miyake et al.,2003)
○ 2000 Tottori-ken Seibu(Ikeda et al.,2002)
△ 2000 Tottori-ken Seibu(Satoh and Irikura,2012)
□ 2005 Fukuoka-ken Seiho-oki(Miyake et al.,2006)
◇ 2005 Fukuoka-ken Seiho-oki(Suzuki and Iwata,2006)
△ 2005 Fukuoka-ken Seiho-oki(Satoh and Kawase,2006)
○ 2005 Fukuoka-ken Seiho-oki(Ikeda et al.,2012)
◇ 2011 Shizuoka-ken Tobu(Somei et al.,2012)

■ 1998 Iwate-ken Hokubu(Miyake et al.,2003)
○ 2004 Niigata-ken Chuetsu(Kamae et al.,2005)
△ 2004 Niigata-ken Chuetsu(Satoh et al.,2007)
▽ 2004 Rumoi(Maeda and Sasatani,2009)
☆ 2007 Noto Hanto(Kurahashi et al.,2008)
▽ 2007 Noto Hanto(Maeda et al.,2008)
○ 2007 Noto Hanto(Ikeda et al.,2011)
☆ 2007 Niigata-ken Chuetsu-oki(Irikura,2008)
○ 2007 Niigata-ken Chuetsu-oki(Shiba,2008)
+ 2007 Niigata-ken Chuetsu-oki(Yamamoto and Takenaka,2009)
○ 2007 Niigata-ken Chuetsu-oki(Kawabe and Kamae,209)
◯ 2008 Iwate-Miyagi Nairiku(Kamae,2008)
☆ 2008 Iwate-Miyagi Nairiku(Irikura and Kurahashi,2008)

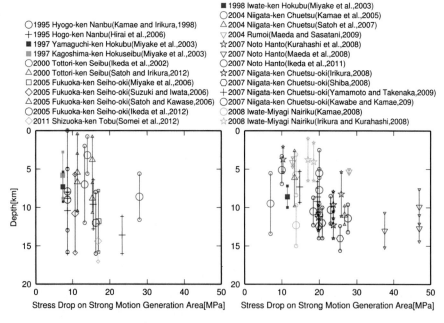

Fig. 6.2 Relations between stress drops and depths of SMGAs for strike-slip faults (*left*) and reverse faults (*right*). *Large symbols* denote the center depth of each SMGA. *Small symbols* denote the top and bottom depths of each SMGA

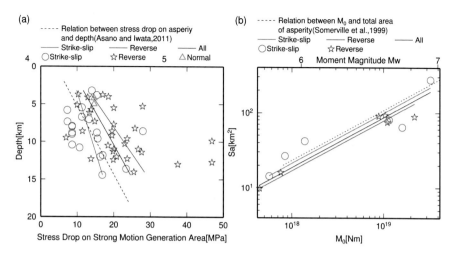

Fig. 6.3 (**a**) Relations between stress drops and the center depths of SMGAs. (**b**) Relations between M_o and total area of SMGAs

stress drop on asperities of 10.5 MPa in the recipe by Irikura and Miyake [3]. The average center depths are 8.90 km, 8.65 km, and 8.65 km for reverse, strike-slip, and all types of faults, respectively. The relations between stress drops $\Delta\sigma_a$ [MPa] on SMGAs and the center depths h [km] are written as

$$\Delta\sigma_a = 0.63h + 7.88 \quad (\text{standard error} = 5.26) \quad \text{for strike-slip faults,} \quad (6.1)$$

$$\Delta\sigma_a = 1.42h + 8.54 \quad (\text{standard error} = 8.39) \quad \text{for reverse faults,} \quad (6.2)$$

$$\Delta\sigma_a = 1.15h + 7.98 \quad (\text{standard error} = 8.05) \quad \text{for all types of faults} \quad (6.3)$$

The depth dependency for reverse faults is stronger than that for strike-slip faults. The stress drops for reverse faults is larger than those for strike-slip faults at the same depth. Although the standard errors [MPa] of the empirical relations are large, Eq. (6.3) means that stress drops increase by about 1 MPa every 1 km in depth. The empirical relations between stress drops on asperities $\Delta\sigma_{asp}$ and the center depths derived by Asano and Iwata [5] for crustal earthquakes in Japan shown in Fig. 6.3a is

$$\Delta\sigma_{asp} = 1.10h + 4.2 \quad (\text{standard error} = 7.2) \quad \text{for all types of faults} \quad (6.4)$$

The depth dependency of the equations of (6.3) and (6.4) are similar, and the absolute value of the stress drop on SMGAs is about 4 MPa larger than the stress drop on asperities.

We also derive the relations between seismic moment M_o [Nm] and total area of SMGAs S_a [km^2] as shown in Fig. 6.3b in which S_a is the average for each earthquake. The equations derived by constraining the slop to be 1/3 are written as

$$S_a = 4.57 \times 10^{-16} \left(M_o \times 10^7\right)^{2/3} \quad \text{(common logarithm of standard error} = 0.18)$$

for strike-slip faults   (6.5)

$$S_a = 3.64 \times 10^{-16} \left(M_o \times 10^7\right)^{2/3} \quad \text{(common logarithm of standard error} = 0.09)$$

for reverse faults   (6.6)

$$S_a = 4.02 \times 10^{-16} \left(M_o \times 10^7\right)^{2/3} \quad \text{(common logarithm of standard error} = 0.15)$$

for all types of faults   (6.7)

The scaling law for total area of asperities S_{asp} by Somerville et al. [4] shown in Fig. 6.3b is written as

$$S_{asp} = 5.00 \times 10^{-16} \left(M_o \times 10^7\right)^{2/3} \tag{6.8}$$

S_a for strike-slip, reverse, and all types of faults are about 0.91, 0.73, and 0.80 times of S_{asp} by Somerville et al. [4]. Although the standard error is large, S_a for each reverse fault is smaller than S_{asp} by Somerville et al. [4]. SMGAs are source models for strong motions in the period range from 0.1 to 5 s, while the asperities are source models for strong motions in the period longer than about 2 s. Therefore, the result that total area of SMGAs is smaller than total area of asperities is interpreted by frequency-dependent source radiations [35].

Short-period spectral level A which means the flat level of acceleration source spectrum [36] is proportional to stress drop and square root of total area of SMGAs (or asperities). Considering the equations of (6.1), (6.2), (6.5), and (6.6), A for reverse faults is larger than A for strike-slip faults. Satoh [35] showed the same results from strong motion records for big crustal earthquakes in Japan using the spectral inversion method. McGarr [37] showed that peak ground velocities PGVs normalized by $M_o^{1/3}$ and hypocentral distances depend on focal depths and are larger for reverse faults than normal faults. He pointed out that these results are expected from frictional laws. In addition he pointed out that data of strike-slip faults were insufficient in his analysis, but the normalized PGVs for strike-slip faults would lie between those for reverse and normal faults. Our results are qualitatively consistent with McGarr's results, although site effects were not considered in McGarr's results.

6.4   Conclusions

We developed empirical relations between stress drops on SMGAs and depths of SMGAs based on previous broadband source models estimated by the empirical Green's function method. A total of 25 source models for 13 crustal earthquakes of Mw from 5.7 to 6.9 in Japan are used in this study. As a result it is found that stress drops on SMGAs for reverse faults are larger than those for strike-slip faults on average. The average stress drops are 21.2 MPa, 13.3 MPa, and 18.0 MPa for reverse, strike-slip, and all types of faults, respectively. In the derived empirical relation for all types of faults, the stress drops increase by about 1 MPa every 1 km in depth. The depth dependency of stress drops for reverse faults is stronger than that for strike-slip faults. We also showed that the total area of SMGAs is about 0.8 times of the total area of asperities by Somerville et al. [4]. This result can be interpreted by frequency-dependent source radiations, since asperities are estimated from longer-period strong motions (>2 s) than SMGAs. The empirical relations derived in this study would be useful for advancement of strong motion predictions for crustal earthquakes by considering together with regional differences and uncertainties.

Acknowledgments   We use focal mechanisms estimated by the Global CMT Project and F-net by NIED and epicenters estimated by JMA. Figures are drawn by GMT [38]. This study is a part of cooperative research of 12 electric power companies.

References

1. Kamae K, Irikura K (1998) Source model of the 1995 Hyogo-ken Nanbu earthquake and simulation of near-source ground motion. Bull Seismol Soc Am 88:400–412
2. Irikura K (1986) Prediction of strong acceleration motions using empirical Green's function. In: Proceedings of the 7th Japan earthquake engineering symposium, Tokyo, Japan, 10–12 Dec 1986, pp 151–156
3. Irikura K, Miyake H (2011) Recipe for predicting strong ground motion from crustal earthquake scenarios. Pure Appl Geophys 168:85–104. doi:10.1007/s00024-010-0150
4. Somerville PG, Irikura K, Graves R, Sawada S, Wald D, Abrahamson D, Iwasaki Y, Kagawa T, Smith N, Kowada A (1999) Characterizing crustal earthquake slip models for the prediction of strong ground motion. Seismol Res Lett 70:59–80
5. Asano K, Iwata T (2011) Characterization of stress drops on asperities estimated from the heterogeneous kinematic slip model for strong motion prediction for inland crustal earthquakes in Japan. Pure Appl Geophys 168:105–116
6. Miyake H, Iwata T, Irikura K (2003) Source characterization for broadband ground-motion simulation: Kinematic heterogeneous source model and strong motion generation area. Bull Seismol Soc Am 93:2531–2545

7. Sekiguchi H, Irikura K, Iwata T (2002) Source inversion for estimating the continuous slip distribution on a fault – introduction of Green's functions convolved with a correction function to give moving dislocation effects in subfaults. Geophys J Int 150:377–391

8. Kuge K, Iwata T, Irikura K (1997) Automatic estimation of earthquake source parameters using waveform data from the K-NET, Programme and Abstracts, the Seismological Society of Japan B16 (in Japanese)

9. Iwata T, Sekiguchi H (2002) Source model of the 2000 Tottori-ken Seibu earthquake and near-source strong ground motion. In: Proceedings of the 11th Japan earthquake engineering symposium, Tokyo, Japan, 20–22 Nov 2002, pp 125–128 (in Japanese with English abstract)

10. Hikima K, Koketsu K (2005) Rupture processes of the 2004 Chuetsu (mid-Niigata prefecture) earthquake, Japan: a series of events in a complex fault system. Geophys Res Lett 32, L18303

11. Asano K, Iwata T (2006) Source process and near-source ground motions of the 2005 West Off Fukuoka Prefecture earthquake. Earth Planets Space 58:93–98

12. Horikawa H (2008) Characterization of the 2007 Noto Hanto, Japan, earthquake. Earth Planets Space 60:1017–1022

13. Shiba Y (2008) Source process and broadband strong motions during the Niigata-ken Chuetsu-oki earthquake in 2007, CRIEPI Research Report N08007:1–21 (in Japanese with English abstract)

14. Horikawa H (2009) Source process of the 2008 Iwate-Miyagi Nairiku earthquake as deduced from strong motion data (preliminary report), Report of CCEP 81:139–141 (in Japanese)

15. Hikima K (2012) Rupture process of the April 11, 2011 Fukushima Hamadori earthquake (Mj 7.0) – two fault planes inferred from strong motion and relocated aftershocks. Zisin, Second Series 64:243–256

16. Hirai T, Kamae K, Naganuma T, Ito S, Nishioka T, Irikura K (2006) Simulation of strong ground motion due to the 1995 Hyogo-ken-Nanbu Earthquake by using characterized source model with branch fault. J JAEE 6(3):1–11 (in Japanese with English abstract)

17. Ikeda T, Kamae K, Miwa S, Irikura K (2002) Source characterization and strong ground motion simulation of the 2000 Tottori-ken Seibu earthquake using the empirical Green's function method. J Struct Constr Eng (Trans Archit Inst Japan) 561:37–45 (in Japanese with English abstract)

18. Satoh T, Irikura K (2012) Short-period spectral level and the total area of strong motion generation area considering type-of-earthquakes and style-of-faultings, Programme and Abstracts, the Seismological Society of Japan:253 (in Japanese)

19. Kamae K, Ikeda T, Miwa S (2005) Source model composed of asperities for the 2004 Mid Niigata Prefecture, Japan, earthquake ($M_{JMA} = 6.8$) by the forward modeling using the empirical Green's function. Earth Planets Space 57:533–538

20. Satoh T, Hijikata K, Uetake T, Tokumitsu R, Dan K (2007) Cause of large peak ground acceleration of the 2004 Niigata-ken Chuetsu earthquake by broadband source inversion Part2. Middle and short-period source inversion, Summaries of technical papers of Annual Meeting, AIJ, B-2:365–366 (in Japanese)

21. Maeda T, Sasatani T (2004) Strong ground motions from an Mj 6.1 inland crustal earthquake in Hokkaido, Japan: the 2004 Rumoi earthquake. Earth Planets Space 61:689–701

22. Suzuki W, Iwata T (2006) Source model of the 2005 west off Fukuoka prefecture earthquake estimated from the empirical Green's function simulation of broadband strong motions. Earth Planets Space 58:99–104

23. Miyake H, Tanaka Y, Sakaue M, Koketsu K, Ishigaki Y (2006) Empirical Green's function simulation of broadband ground motions on Genkai Island during the 2005 West Off Fukuoka prefecture earthquake. Earth Planets Space 58:1637–1642

24. Satoh T, Kawase H (2006) Estimation of characteristic source model of the 2005 west off Fukuoka prefecture earthquake based on empirical Green's function method. In: Proceedings of the 12th Japan earthquake engineering symposium, Tokyo, Japan, 3–5 Nov 2006, pp 170–173 (in Japanese with English abstract)

25. Ikeda T, Kamae K, Irikura K (2011) Source modeling using the empirical Green's function method and strong ground motion estimation considering nonlinear site effect: an application to the 2005 west off Fukuoka prefecture earthquake and the 2007 Noto Hanto earthquake. J Struct Constr Eng (Trans Archit Inst Japan) 665:1253–1261 (in Japanese with English abstract)
26. Kurahashi S, Masaki K, Irikura K (2008) Source model of the 2007 Noto-Hanto earthquake (Mw 6.7) for estimating broad-band strong ground motion. Earth Planets Space 60:89–94
27. Maeda T, Ichiyanagi M, Takahashi H, Honda R, Yamaguchi T, Kasahara M, Sasatani T (2008) Source parameters of the 2007 Noto Hanto earthquake sequence derived from strong motion records at temporary and permanent stations. Earth Planets Space 60:1011–1016
28. Irikura K (2008) http://www.kojiro-irikura.jp/pdf/chikyuokangaerukai.pdf. Accessed 5 Jan 2015 (in Japanese)
29. Yamamoto Y, Takenaka H (2009) Source modeling of the 2007 Niigataken Chuetsu oki earthquake using the empirical Green's function method. Zisin, Second Series 62:47–59 (in Japanese with English abstract)
30. Kawabe H, Kamae K (2010) Source modeling and 3D ground motion simulation of the 2007 Niigataken Chuetsu-oki earthquake (Mj6.8). In: Proceedings of the 13th Japan earthquake engineering symposium, Tsukuba, Japan, 17–10 Nov 2010, pp 1899–1906 (in Japanese with English abstract)
31. Kamae K (2008) http://www.rri.kyoto-u.ac.jp/jishin/iwate_miyagi_1.html. Accessed 5 Jan 2015 (in Japanese)
32. Irikura K, Kurahashi S (2008) Modeling of source fault and generation of high-acceleration ground motion for the 2008 Iwate-miyagi-nairiku earthquake, Programme and Abstracts, JSAF Fall Meeting: 8–11, t. Accessed 5 Jan 2015 (in Japanese)
33. Somei K, Miyakoshi K, Kamae K (2012) Source model of the 2011 East Shizuoka prefecture, Japan, earthquake by using the empirical Green's function method. Japan Geoscience Union Meeting, SSS26-P27
34. Somei K, Miyakoshi K, Irikura K (2011) Estimation of source model and strong motion simulation for the 2011 East Fukushima Prefecture Earthquake, Programme and Abstracts, the Seismological Society of Japan, Fall Meeting:211 (in Japanese)
35. Satoh T (2010) Scaling law of short-period source spectra for crustal earthquakes in Japan considering style of faulting of dip-slip and strike-slip. J Struct Constr Eng (Trans Archit Inst Japan) 651:923–932 (in Japanese with English abstract)
36. Dan K, Watanabe M, Sato T, Ishii T (2001) Short-period source spectra inferred from variable-slip rupture models and modeling of earthquake faults for strong motion prediction by semi-empirical method. J Struct Constr Eng (Trans Archit Inst Japan) 545:51–62 (in Japanese with English abstract)
37. McGarr A (1984) Scaling of ground motion parameters, state of stress, and focal depth. J Geophys Res 89:6969–6979. doi:10.1029/JB089iB08p06969
38. Wessel P, Smith WHF (1998) New, improved version of generic mapping tools released. Eos Trans Am Geophys Union 79:579

Chapter 7
Heterogeneous Dynamic Stress Drops on Asperities in Inland Earthquakes Caused by Very Long Faults and Their Application to the Strong Ground Motion Prediction

Kazuo Dan, Masanobu Tohdo, Atsuko Oana, Toru Ishii,
Hiroyuki Fujiwara, and Nobuyuki Morikawa

Abstract We compiled the stress drops on the asperities in inland earthquakes caused by strike-slip faults. Then, we applied the log-normal distribution to the data and obtained the medium of 10.7 MPa and the logarithmic standard deviation of 0.45. Also, we compiled the stress drops on the asperities in inland earthquakes caused by reverse faults and obtained the medium of 17.1 MPa and the logarithmic standard deviation of 0.39.

By using the obtained log-normal distributions, we examined a procedure for assigning the heterogeneous dynamic stress drops to each asperity. We adopted 12.2 MPa, which had been estimated by Dan et al. (J Struct Constr Eng (Trans Archit Inst Japan), 76:(670):2041–2050, 2011) for long strike-slip faults, as the medium, and 18.7 MPa, which had been estimated by Dan et al. (J Struct Constr Eng (Trans Archit Inst Japan), 80(707):47–57, 2015) for long reverse faults.

Moreover, we truncated the log-normal distributions of the dynamic stress drops on the asperities at the value of 3.4 MPa for strike-slip faults and of 2.4 MPa for reverse faults because they should be larger than the dynamic stress drop averaged over the entire fault.

Finally, we proposed a procedure for evaluating fault parameters taking into account of the heterogeneous dynamic stress drops on the asperities and calculated strong ground motions. The results had wider variations of the peak ground accelerations and velocities than those with uniform dynamic stress drops on the asperities, while the averages were almost the same.

K. Dan (✉) • A. Oana • T. Ishii
Institute of Technology, Shimizu Corporation, Tokyo, Japan
e-mail: kazuo.dan@shimz.co.jp

M. Tohdo
Ohsaki Research Institute, Inc., Chiyoda-ku, Tokyo, Japan

H. Fujiwara • N. Morikawa
National Research Institute for Earth Science and Disaster Prevention, Tokyo, Japan

© The Author(s) 2016
K. Kamae (ed.), *Earthquakes, Tsunamis and Nuclear Risks*,
DOI 10.1007/978-4-431-55822-4_7

Keywords Strong motion prediction • Inland earthquake • Long fault • Fault parameter • Asperity • Dynamic stress drop • Heterogeneity

7.1 Introduction

Dan et al. [1, 2] proposed a procedure for evaluating the parameters of long strike-slip faults, evaluated fault parameters based on the proposed procedure, and calculated strong ground motions. Also, Dan et al. [3] carried out the same study for long reverse faults. In these studies, they treated the dynamic stress drops on the asperities as the uniform ones. But, it is hard to assume that all the dynamic stress drops on the asperities would be uniform in the actual earthquakes. Especially, in long faults, the number of the asperities is thought to be large, and the heterogeneity of the dynamic stress drops is easier to be observed. This tendency should cause large effects on the spatial distribution of the predicted strong motions.

Hence, in this paper, we proposed a procedure for evaluating fault parameters taking into account the heterogeneous dynamic stress drops on the asperities and calculated strong ground motions to compare the results with ground motion prediction equations by Si and Midorikawa [4] and with the results with uniform dynamic stress drops on the asperities.

7.2 Statistics of the Heterogeneous Stress Drops on the Asperities

7.2.1 Strike-Slip Faults

At first, we compiled heterogeneous stress drops on the asperities in the past inland earthquakes caused by strike-slip faults. Table 7.1 shows stress drops on the asperities or SMGAs (strong motion generation areas) in the past earthquakes obtained by previous studies [5–9]. In Table 7.1, when the stress drops on the asperities in one earthquake were different from each other, each value was adopted independently, but when all the values of the stress drops on the asperities in one earthquake

Table 7.1 Stress drops on the asperities in the past inland earthquakes caused by strike-slip faults

Earthquakes	References	Stress drops on the asperities (MPa)		
		$\Delta\sigma_{asp1}$	$\Delta\sigma_{asp2}$	$\Delta\sigma_{asp3}$
1995 Hyogo-Ken Nanbu	Kamae and Irikura [5]	8.6	16.3	8.6
1999 Kocaeli, Turkey	Kamae and Irikura [6]	12.0	5.0	10.0
2000 Tottori-ken Seibu	Ikeda et al. [7]	28.0	14.0	–
″	Muto et al. [8]	8.7	7.3	–
2005 Fukuoka-ken Seiho-oki	Satoh and Kawase [9]	11.3	11.3	–

Fig. 7.1 Fitting of the log-normal distribution to the stress drops on the asperities in the past inland earthquakes caused by strike-slip faults

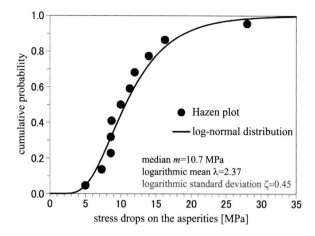

were the same, that value was adopted as one data. Here, the stress drop $\Delta\sigma$, also called static stress drop, is the difference between the initial shear stress on the fault before the earthquake and final shear stress after the earthquake at the time all the rupture on the fault terminates and the stress status becomes stable, and the dynamic stress drop $\Delta\sigma^{\#}$ is the difference between the initial shear stress and the shear stress at the time the rupture terminates at a certain point on the fault while the rupture may not terminate at other points. Although the stress drop $\Delta\sigma$ and the dynamic stress drop $\Delta\sigma^{\#}$ are different from each other in general, we assumed the difference to be negligible in this paper.

When the number of the stress drops was K, we adopted Hazen plot and assigned a cumulative probability (non-exceeding probability) P_k as follows:

$$P_k = 1 - \frac{k - 0.5}{K}. \tag{7.1}$$

We calculated a logarithmic mean and a logarithmic standard deviation of the data and fitted a log-normal distribution to the data. Figure 7.1 shows the result. The logarithmic mean of the stress drops was calculated to be 2.37 (the median = 10.7 MPa) and the logarithmic standard deviation to be 0.45.

7.2.2 Reverse Faults

Next, we compiled heterogeneous stress drops on the asperities in the past inland earthquakes caused by reverse faults.

Table 7.2 shows stress drops on the asperities or SMGAs (strong motion generation areas) in the past earthquakes obtained by previous studies [6, 10–18].

We calculated a logarithmic mean and a logarithmic standard deviation of the data and fitted a log-normal distribution to the data. Figure 7.2 shows the result.

Table 7.2 Stress drops on the asperities in the past inland earthquakes caused by reverse faults

Earthquakes	References	Stress drops on the asperities (MPa)			
		$\Delta\sigma_{asp1}$	$\Delta\sigma_{asp2}$	$\Delta\sigma_{asp3}$	$\Delta\sigma_{asp4}$
1999 Chi-Chi, Taiwan	Kamae and Irikura [6]	10.0	10.0	10.0	–
2004 Niigata-Chuetsu	Kamae et al. [10]	7.0	20.0	–	–
//	Satoh et al. [11]	26.7	13.4	–	–
2007 Noto-Hanto	Kamae et al. [12]	20.0	20.0	10.0	–
//	Kurahasi et al. [13]	25.8	10.3	–	–
2007 Niigata-Chuetsu-oki	Irikura et al. [14]	23.7	23.7	19.8	–
//	Kamae and Kawabe [15]	18.4	27.6	27.6	–
2008 Iwate-Miyagi Nairiku	Kamae [16]	13.8	13.8	–	–
//	Irikura and Kurahashi [17]	17.0	18.5	–	–
2008 Wenchuan, China	Irikura and Kurahashi [18]	13.2	13.2	13.2	13.2

Fig. 7.2 Fitting of the log-normal distribution to the stress drops on the asperities in the past inland earthquakes caused by reverse faults

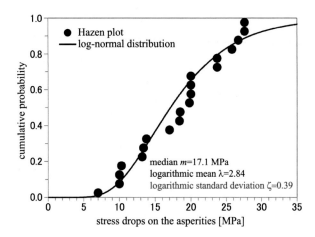

The logarithmic mean of the stress drops was calculated to be 2.84 (the median = 17.1 MPa) and the logarithmic standard deviation to be 0.39.

7.3 Procedure for Evaluating Fault Parameters

We examined how to assign the heterogeneous stress drops to each asperity based on the cumulative probability distribution obtained in Sect. 7.2 for the strong motion prediction.

The median of the stress drops on the asperities in strike-slip faults is consistent with the value of 12.2 MPa estimated by Dan et al. [1] as the geometrical mean of the dynamic stress drops on the asperities in strike-slip faults and that for reverse faults is consistent with the value of 18.7 MPa estimated by Dan et al. [3] as the geometrical mean of the dynamic stress drops on asperities in reverse faults. Hence, we adopted 12.2 MPa as the median for strike-slip faults and 18.7 MPa for reverse faults. As for the variation, we adopted 0.45 as the logarithmic standard deviations for strike-slip faults as shown in Fig. 7.1 and 0.39 for reverse faults as shown in Fig. 7.2.

Figure 7.3 shows the cumulative probability function for strike-slip faults indicated by the red line. Here, we truncated the function less than 3.4 MPa because the averaged dynamic stress drop on the entire fault is 3.4 MPa [1]. For reverse faults, we truncated the function less than 2.4 MPa because the averaged dynamic stress drop on the entire fault is 2.4 MPa [3].

We chose the stress drop at the middle point in the line divided equally of the vertical axis for the cumulative probability function, as shown in Fig. 7.3 based on the idea of Hazen plot, and assigned it to the dynamic stress drop on each asperity.

When we apply the heterogeneity to the dynamic stress drops on the asperities, the seismic moment and the short-period level would become different from those of the original fault model with the uniform dynamic stress drops on the asperities. Here, the short-period level is the flat level of the acceleration source spectrum in the short-period range. Also, the relationship would not be preserved that the slips on the asperities should be proportional to the dynamic stress drops on the asperities and the equivalent radii of the asperities if the slips on the asperities obey the similar trend as the equation of the constant stress drop on a circular crack. Hence, we preserved the seismic moment by adjusting the areas of the asperities so that the averaged dynamic stress drop on the entire fault should be preserved. Also, we reevaluated the slips on the asperities so that the relationship was preserved that the slips on the asperities should be proportional to the dynamic stress drops on the asperities and the equivalent radii of the asperities.

Because it is impossible to preserve the short-period level, we just confirmed that the value of the short-period level did not change largely. In this paper, we assigned the heterogeneous dynamic stress drops to the asperities randomly.

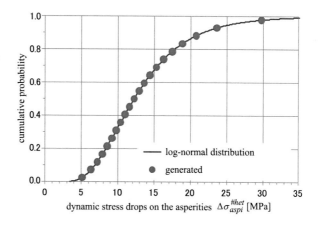

Fig. 7.3 Generated examples of the heterogeneous dynamic stress drops on the asperities in the strike-slip fault ($N = 21$)

When we evaluate fault parameters with heterogeneous dynamic stress drops, we first assume the parameters with the uniform dynamic stress drop such as the areas S_{aspi} of the i-th asperity, the dynamic stress drop $\Delta\sigma^{\#}_{\mathrm{asp}}$, and the averaged slip D_{asp} on the asperities.

We put the heterogeneous stress drop on the asperity as $\Delta\sigma^{\#\,\mathrm{het}}_{\mathrm{aspi}}$, which is generated by the way of Fig. 7.3. When we write the area of the i-th asperity with heterogeneous dynamic stress drop as $S^{\mathrm{het}}_{\mathrm{aspi}}$ and its ratio to S_{aspi} as p as follows:

$$S^{\mathrm{het}}_{\mathrm{aspi}} = p S_{\mathrm{aspi}}, \tag{7.2}$$

we obtain

$$\left(\sum_i \Delta\sigma^{\#\,\mathrm{het}}_{\mathrm{aspi}} S^{\mathrm{het}}_{\mathrm{aspi}}\right)\bigg/ S_{\mathrm{seis}} = \left(p\sum_i \Delta\sigma^{\#\,\mathrm{het}}_{\mathrm{aspi}} S_{\mathrm{aspi}}\right)\bigg/ S_{\mathrm{seis}}$$

$$= \left(\sum_i \Delta\sigma^{\#}_{\mathrm{asp}} S_{\mathrm{aspi}}\right)\bigg/ S_{\mathrm{seis}}, \tag{7.3}$$

because the averaged dynamic stress drop on the entire fault S_{seis} should be preserved.

Then, we obtain

$$p = \left(\sum_i \Delta\sigma^{\#}_{\mathrm{asp}} S_{\mathrm{aspi}}\right)\bigg/ \left(\sum_i \Delta\sigma^{\#\,\mathrm{het}}_{\mathrm{aspi}} S_{\mathrm{aspi}}\right). \tag{7.4}$$

The $S^{\mathrm{het}}_{\mathrm{aspi}}$ can be evaluated by substituting p of Eq. (7.4) for Eq. (7.2).

On the other hand, the slip $D^{\mathrm{het}}_{\mathrm{aspi}}$ on the asperity should be proportional to the dynamic stress drop on the asperity and the equivalent radius of the asperity. Hence, if we write the proportionality constant as q, the $D^{\mathrm{het}}_{\mathrm{aspi}}$ can be written by

$$D^{\mathrm{het}}_{\mathrm{aspi}} = q\Delta\sigma^{\#\,\mathrm{het}}_{\mathrm{aspi}} \sqrt{S^{\mathrm{het}}_{\mathrm{aspi}}/\pi}. \tag{7.5}$$

The averaged slip D_{asp} on the asperities is obtained by

$$D_{\mathrm{asp}} = \left(\sum_i D^{\mathrm{het}}_{\mathrm{aspi}} S^{\mathrm{het}}_{\mathrm{aspi}}\right)\bigg/ \sum_i S^{\mathrm{het}}_{\mathrm{aspi}}$$

$$= \left(q\sum_i \Delta\sigma^{\#\,\mathrm{het}}_{\mathrm{aspi}} S^{\mathrm{het}}_{\mathrm{aspi}} \sqrt{S^{\mathrm{het}}_{\mathrm{aspi}}/\pi}\right)\bigg/ \sum_i S^{\mathrm{het}}_{\mathrm{aspi}}. \tag{7.6}$$

Then, we obtain

$$q = D_{asp} \sum_i S_{aspi}^{het} \Bigg/ \left(\sum_i \Delta\sigma_{aspi}^{\#het} S_{aspi}^{het} \sqrt{S_{aspi}^{het}/\pi} \right). \qquad (7.7)$$

The D_{aspi}^{het} can be evaluated by substituting q of Eq. (7.7) for Eq. (7.5). In addition, in the case that the D_{aspi}^{het} is not larger than the averaged slip D on the entire fault, we should assign again the heterogeneous dynamic stress drops to the asperities randomly.

It is also necessary to evaluate the parameters for the background, and we can adopt the same procedure as those by Dan et al. [1] and Dan et al. [3] for evaluating the parameters of the fault model with the uniform dynamic stress drops on the asperities.

Figure 7.4 shows the evaluation procedure of the fault parameters mentioned above.

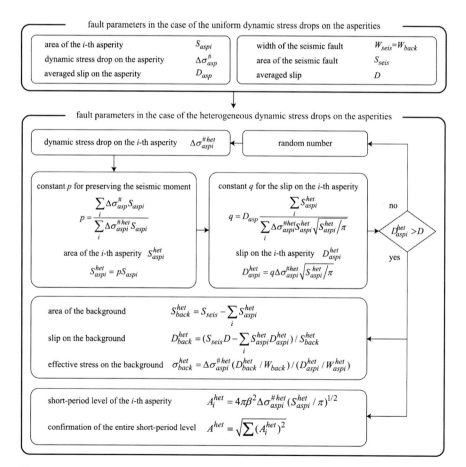

Fig. 7.4 Evaluation procedure of the fault parameters in the case of the heterogeneous dynamic stress drops on the asperities

7.4 Examples of Strong Motion Prediction Under Heterogeneous Dynamic Stress Drops on the Asperities

For the active strike-slip fault 360 km long along the Median Tectonic Line, Japan, shown in Fig. 7.5, we made two models with the uniform dynamic stress drops on the asperities and with the heterogeneous dynamic stress drops.

Figure 7.6 shows the fault model with the uniform dynamic stress drops on the 21 asperities. Figure 7.7 shows the fault model with the heterogeneous dynamic stress drops evaluated by the procedure in Fig. 7.4. We confirmed that the short-period level of the model in Fig. 7.7 was 10 % larger than that of the model in Fig. 7.6.

Next, we calculated strong ground motions at 10-km-mesh points around the faults by the stochastic Green's function method [19].

Figure 7.8 compares the peak ground accelerations and velocities for the model with the uniform dynamic stress drops on the asperities and the ground motion prediction equations by Si and Midorikawa [4]. Figure 7.9 compares the peak ground accelerations and velocities for the model with heterogeneous dynamic stress drops and the ground motion prediction equations by Si and Midorikawa [4]. We find that the accelerations and velocities in Fig. 7.9 have larger deviation than those in Fig. 7.8. Especially, in the vicinity of the fault trace, while most of the peak ground accelerations and velocities for the model with the uniform dynamic stress drops are within the mean plus/minus standard deviation of the ground

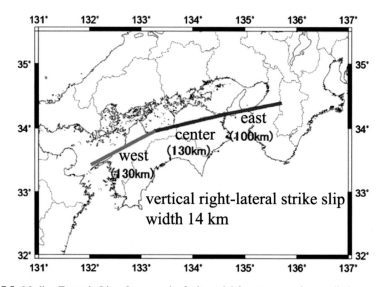

Fig. 7.5 Median Tectonic Line, Japan, and a fault model for strong motion prediction

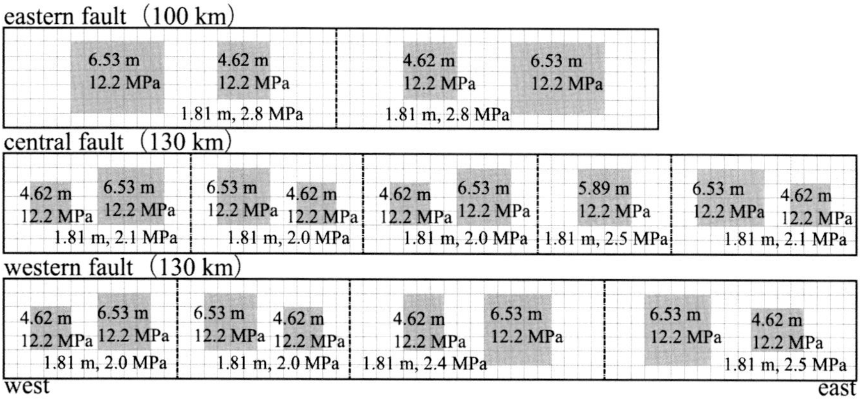

eastern fault （100 km）

6.53 m 12.2 MPa	4.62 m 12.2 MPa	4.62 m 12.2 MPa	6.53 m 12.2 MPa
	1.81 m, 2.8 MPa	1.81 m, 2.8 MPa	

central fault （130 km）

4.62 m 12.2 MPa	6.53 m 12.2 MPa	6.53 m 12.2 MPa	4.62 m 12.2 MPa	4.62 m 12.2 MPa	6.53 m 12.2 MPa	5.89 m 12.2 MPa	6.53 m 12.2 MPa	4.62 m 12.2 MPa
1.81 m, 2.1 MPa		1.81 m, 2.0 MPa		1.81 m, 2.0 MPa	1.81 m, 2.5 MPa		1.81 m, 2.1 MPa	

western fault （130 km）

4.62 m 12.2 MPa	6.53 m 12.2 MPa	6.53 m 12.2 MPa	4.62 m 12.2 MPa	4.62 m 12.2 MPa	6.53 m 12.2 MPa	6.53 m 12.2 MPa	4.62 m 12.2 MPa
1.81 m, 2.0 MPa		1.81 m, 2.0 MPa		1.81 m, 2.4 MPa			1.81 m, 2.5 MPa

west   east

Fig. 7.6 Fault model with uniform dynamic stress drops on the asperities

eastern fault （100 km）

7.13 m 15.3 MPa	3.42 m 10.4 MPa	8.14 m 23.6 MPa	4.77 m 9.8 MPa
	1.84 m, 3.2 MPa	1.84 m, 3.0 MPa	

central fault （130 km）

4.26 m 11.6 MPa	6.71 m 12.9 MPa	4.59 m 7.2 MPa	8.49 m 18.9 MPa	8.70 m 17.4 MPa	4.48 m 6.4 MPa	5.89 m 5.1 MPa	5.74 m 13.6 MPa	6.19 m 20.8 MPa
1.84 m, 2.1 MPa		1.84 m, 1.7 MPa		1.84 m, 1.5 MPa		1.84 m, 1.0 MPa	1.84 m, 2.6 MPa	

western fault （130 km）

5.20 m 14.4 MPa	6.24 m 12.2 MPa	3.97 m 8.6 MPa	9.74 m 29.8 MPa	5.14 m 9.2 MPa	6.27 m 7.9 MPa	7.14 m 16.3 MPa	3.40 m 11.0 MPa
1.84 m, 2.1 MPa		1.84 m, 2.3 MPa		1.84 m, 1.6 MPa			1.84 m, 3.0 MPa

west   east

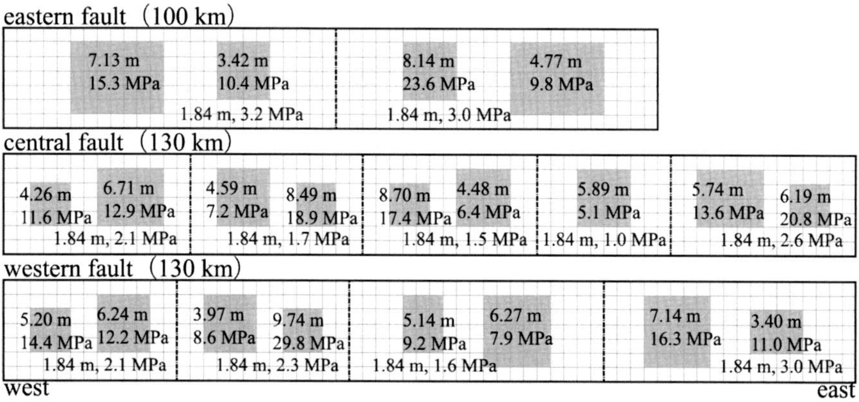

Fig. 7.7 Fault model with heterogeneous dynamic stress drops on the asperities

motion prediction equations by Si and Midorikawa [4], some of those for the model with the heterogeneous dynamic stress drops are beyond the mean plus standard deviation. However, the averages are almost the same.

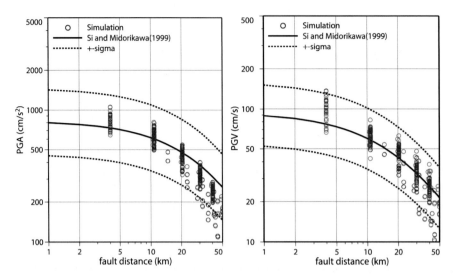

Fig. 7.8 Comparison between the strong motions predicted by the fault model with uniform dynamic stress drops on the asperities and ground motion prediction equations by Si and Midorikawa [4]

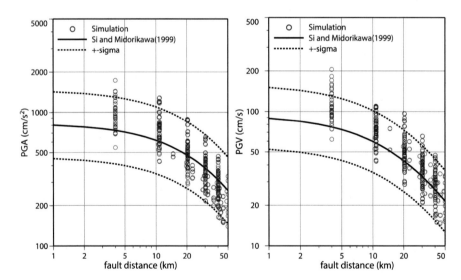

Fig. 7.9 Comparison between the strong motions predicted by the fault model with heterogeneous dynamic stress drops on the asperities and ground motion prediction equations by Si and Midorikawa [4]

7.5   Conclusions

We proposed a procedure for evaluating fault parameters taking into account of the heterogeneous dynamic stress drops on the asperities and calculated strong ground motions. The results had wider variations of the peak ground accelerations and velocities than those with the uniform dynamic stress drops on the asperities, while the averages were almost the same.

The procedure proposed in this paper can be applied not only to very long faults of M_W 8 class earthquakes but also to medium-sized faults of M_W 7 class earthquakes.

Acknowledgments   This study is a part of the examination results in Research Project on Seismic Hazard Assessment for Japan by the National Research Institute for Earth Science and Disaster Prevention.

References

1. Dan K, Ju D, Irie K, Arzpeima S, Ishii Y (2011) Estimation of averaged dynamic stress drops of inland earthquakes caused by long strike-slip faults and its application to asperity models for predicting strong ground motions. J Struct Constr Eng (Trans Archit Inst Japan) 76(670): 2041–2050
2. Dan K, Ju D, Shimazu N, Irie K (2012) Strong ground motions simulated by asperity models based on averaged dynamic stress drops on inland earthquakes caused by long strike-slip faults. J Struct Constr Eng (Trans Archit Inst Japan) 77(678):1257–1264
3. Dan K, Irie K, Ju D, Shimazu N, Torita H (2015) Procedure for estimating parameters of fault models of inland earthquakes caused by long reverse faults. J Struct Constr Eng (Trans Archit Inst Japan) 80(707):47–57
4. Si H, Midorikawa S (1999) New attenuation relationships for peak ground acceleration and velocity considering effects of fault type and site condition. J Struct Constr Eng (Trans Archit Inst Japan) (523):63–70
5. Kamae K, Irikura K (1997) A fault model of the 1995 Hyogo-Ken Nanbu earthquakes and simulation of strong ground motion in near-source area. J Struct Constr Eng (Trans Archit Inst Japan) (500):29–36
6. Kamae K, Irikura K (2002) Source characterization and strong motion simulation for the 1999 Kocaeli, Turkey and the 1999 Chi-Chi, Taiwan earthquakes. In: Proceedings of the 11th Japan earthquake engineering symposium, pp 545–550
7. Ikeda T, Kamae K, Miwa S, Irikura K (2002) Strong ground motion simulation of the 2000 Tottori-Ken Seibu earthquake using the hybrid technique. In: Proceedings of the 11th Japan earthquake engineering symposium, pp 579–582
8. Muto M, Shimazu N, Dan K, Abiru T (2009) Asperity models for the crustal earthquakes in Chugoku District based on short-period level by spectral inversion, Part 2: 2000 Tottori-Ken Seibu earthquake, Summaries of technical papers of Annual Meeting Architectural Institute of Japan, pp 151–152

9. Satoh T, Kawase H (2006) Estimation of characteristic source model of the 2005 west off Fukuoka prefecture earthquake based on empirical Green's function method. In: Proceedings of the 12th Japan earthquake engineering symposium, pp 170–173

10. Kamae K, Ikeda T, Miwa S (2005) Source model composed of asperities for the 2004 Mid Niigata Prefecture, Japan, earthquake ($M_{JMA} = 6.8$) by the forward modeling using the empirical Green's function method. Earth Planets Space 57:533–538

11. Satoh T, Hijikata K, Uetake T, Tokumitsu R, Dan K (2007) Cause of large peak ground acceleration of the 2004 Niigata-Ken Chuetsu earthquake by broadband source inversion, Part 2. Middle and short-period source inversion, Summaries of technical papers of Annual Meeting Architectural Institute of Japan, pp 365–366

12. Kamae K, Ikeda T, Miwa S (2014) http://www.rri.kyoto-u.ac.jp/jishin/eq/notohantou/notohantou.html. Referred on 3 June 2014

13. Kurahashi S, Masaki K, Irikura K (2008) Source model of the 2007 Noto-Hanto earthquake (Mw 6.7) for estimating broad-band strong ground motion. Earth Planets Space 60:89–94

14. Irikura K, Kagawa T, Miyakoshi K, Kurahashi S (2014) http://www.kojiro-irikura.jp/pdf/jishingakkai2009PPT.pdf. Referred on 3 June 2014

15. Kamae K, Kawabe H (2014) http://www.rri.kyoto-u.ac.jp/jishin/eq/niigata_chuetsuoki_5/chuuetsuoki_20080307.pdf. Referred on 3 June 2014

16. Kamae K (2014) http://www.rri.kyoto-u.ac.jp/jishin/iwate_miyagi_1.html. Referred on 3 June 2014

17. Irikura K, Kurahashi S (2014) Modeling of source fault and generation of high-acceleration ground motion for the 2008 Iwate-Miyagi-Nairiku earthquake, Programme and Abstracts, Japanese Society of Active Fault Studies 2008 Fall Meeting. http://www.kojiro-irikura.jp/pdf/katudansougakkai2008.pdf. Referred on 3 June 2014

18. Irikura K, Kurahashi S (2009) "Recipe" of strong motion prediction for great earthquakes with mega fault systems, Programme and Abstracts, The Seismological Society of Japan 2009, Fall Meeting. http://www.kojiro-irikura.jp/pdf/jishingakkai2009PPT.pdf. Referred on 3 June 2014

19. Dan K, Ju D, Muto M (2010) Modeling of subsurface fault for strong motion prediction inferred from short active fault observed on ground surface. J Struct Constr Eng (Trans Archit Inst Japan) 75(648):279–288

Chapter 8
Simulation of Broadband Strong Motion Based on the Empirical Green's Spatial Derivative Method

Michihiro Ohori

Abstract In our previous studies (Ohori and Hisada, Zisin 2(59):133–146, 2006, Bull Seismol Soc Am 101:2872–2886, 2011), we simulated the strong-motion records of the mainshock ($M_J5.4$) of the 2001 Hyogo-ken Hokubu earthquake, Japan, on the basis of the empirical Green's tensor spatial derivative (EGTD) estimated from data of 11 aftershocks ($M_J3.5$–4.7). The agreement between the observed and calculated waveforms at the closest station in source distance was satisfactory over a long duration, and the amplitude was well reproduced. But considering the lowest corner frequency of about 1.0 Hz for the mainshock, we targeted 0.2–1.0 Hz band-pass-filtered velocity waveforms. In the present study, we tried to simulate the broadband strong motions beyond the corner frequencies for the same events as in our previous studies mentioned above. To correct the discrepancies among the corner frequencies of events, we assumed the scaling law based on the ω^{-2} model (Aki, J Geophys Res 72:1217–1231, 1967) and compensated the spectral amplitude decay beyond the corner frequency. After estimating the EGTD from 11 aftershock events using 0.2–10 Hz band-pass-filtered waveforms, we simulated the strong-motion records for the mainshock and aftershocks. In simulation of each event, the EGTD elements were multiplied by the moment tensor elements followed by summation and corrected in the spectral amplitude, taking the corner frequency of each event into account. As example results, the simulated waveforms at the closest epicentral distance was compared with the observed ones. The agreement between the calculated and observed waveforms was acceptable for most of events.

Keywords Empirical Green's tensor spatial derivative • Broadband strong motion • ω^{-2} model • Corner frequency • Source time function • Waveform inversion

M. Ohori (✉)
Research Institute of Nuclear Engineering, University of Fukui, Tsuruga, Japan
e-mail: ohorim@u-fukui.ac.jp

© The Author(s) 2016   99
K. Kamae (ed.), *Earthquakes, Tsunamis and Nuclear Risks*,
DOI 10.1007/978-4-431-55822-4_8

8.1   Introduction

The empirical Green's function (EGF) method, proposed by Hartzell [1] and extended by Irikura [2], has been recognized as one of the most practical techniques to predict the strong idground motion produced by large earthquakes. The use of this method is limited to the case when the focal mechanism of a small event is identical or similar to that of a targeted event. On the other hand, the empirical Green's tensor spatial derivative (EGTD) method, proposed by Plicka and Zahradnik [3], has the potential to deal with the difference in focal mechanisms between small events and a targeted event and predict the ground motion for an event with an arbitrary focal mechanism. The EGTD elements are estimated through a kind of single-station inversion using waveform data for several small events whose focal mechanisms and source time functions are well determined. This technique is expected to provide considerably accurate and stable prediction results, but discussion of its application has been limited in the literature [4–8]. In the previous studies [6, 8], considering the lowest corner frequency of about 1.0 Hz for the mainshock, we targeted 0.2–1.0 Hz band-pass-filtered velocity waveforms. In the present study, we simulate the broadband strong motions between 0.2 and 10 Hz for the same events as in the previous studies [6, 8].

8.1.1   Targeted Events and Stations

In this study, we targeted the mainshock ($M_J5.4$, labeled as "Event 1") and 25 aftershocks ($M_J3.1$–4.7, labeled as "Events 2-26") of the 2001 Hyogo-ken Hokubu earthquake. We used the strong motion records at the target station, HYG004, one of the K-NET stations operated by the NIED (National Research Institute for Earth Science and Disaster Prevention). The source information of these events was determined by the united hypocenter catalog of the JMA (Japan Meteorological Agency). In Fig. 8.1, we show the distribution of focal mechanisms and source time functions as well as location of the station. Details of source parameters can be found in Ohori and Hisada [8]. Data for the mainshock and 15 of the aftershocks ($M_J3.5$–4.7) were recorded at HYG004. As one of the K-NET stations, HYG004 was chosen as the target station because of the data quality. It is on a rock site located at the closest epicentral distance (6–10 km) from the fault zone, whose range was 4 km in the east–west direction and 6 km in the north–south direction. The observed acceleration records at HYG004 for the mainshock and 15 aftershocks were integrated into velocity waveform data with a band-pass filter of 0.2–10 Hz. To enhance the accuracy of simulation by the EGTD method, it is desirable to know the focal mechanisms and source time functions accurately as much as possible. In this study, we used the source model which was reevaluated in our previous work [6]. Among 15 events, 4 aftershock events, 3, 7, 19, 26, are

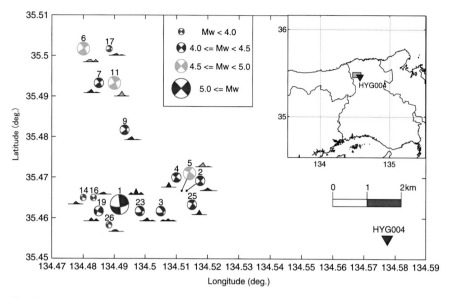

Fig. 8.1 Map showing the focal mechanisms and source time functions of 16 events (mainshock and 15 aftershocks) as well as location of the station HYG004 used in present study

excluded in the following EGTD inversion because of the relatively large discrepancy in waveform matching between the observation data and synthesis.

8.1.1.1 Estimation of the EGTD

The estimation method of the EGTD has been fully explained by Ohori and Hisada [6, 8]. It is applicable to simulation of the strong motion in a frequency range below the corner frequency. Hereafter, we summarize briefly the method and provide additional explanations on how to compensate the spectral amplitude decay beyond the corner frequencies and how to simulate the broadband ground motion.

8.1.1.2 Basic Equations

Ground motion displacement $u_i(x_o, t)$ $(i = x,y,z)$ excited by a double-couple point source is theoretically expressed as the convolution of moment tensor elements $M_{pq}(x_s,\tau)$ $(p,q = x,y,z)$ and Green's tensor spatial derivative elements $G_{ip,q}(x_o, t \mid x_s,\tau)$

$$u_i(x_o,t) = M_{pq}(x_s,\tau) \cdot G_{ip,q}(x_o,t \mid x_s,\tau) \tag{8.1}$$

Hereafter, we abbreviate $u_i(x_o, t)$, $M_{pq}(x_s,\tau)$, and $G_{ip,q}(x_o, t \mid x_s,\tau)$ as u_i, M_{pq}, and $G_{ip,q}$.

Explicit expressions of M_{pq} for a double-couple point source are found in the literature (e.g., Aki and Richards [9]). Considering symmetrical conditions $(M_{pq} = M_{qp})$ and no volume change $[(M_{xx} = -(M_{yy} + M_{zz})]$ of the moment tensor elements, we can rewrite Eq. (8.1) as

$$u_i = \sum_{j=1}^{5} M_j \cdot G_{ij} \tag{8.2}$$

where M_j ($j = 1,2,\ldots,5$) is defined by $M_1 = M_{xy}$, $M_2 = M_{yy}$, $M_3 = M_{yz}$, $M_4 = M_{yz}$, $M_5 = M_{zz}$, and G_{ij} ($j = 1,2,\ldots,5$) is defined by $G_{i1} = G_{ix,y} + G_{iy,x}$, $G_{i2} = G_{iy,y} - G_{x,x}$, $G_{i3} = G_{iy,z} + G_{iz,y}$, $G_{i4} = G_{ix,z} + G_{iz,x}$, $G_{i5} = G_{iz,z} - G_{ix,x}$. In the moment tensor inversion, u_i and G_{ij} are given and M_j are the unknowns to be solved in a least-squares sense. Conversely, in the EGTD inversion, u_i and M_j are given and G_{ij} are the unknowns to be solved. Note that the EGTD inversion is carried out for each component at each station using data from several events simultaneously, whereas the moment tensor inversion is done for a particular event using data of all possible components at all possible stations simultaneously. It should be emphasized that the moment tensor elements are determined by the source parameters and the Green's tensor spatial derivative elements are by the underground structure of the area surrounding the source and the station.

8.1.1.3 Correction of the Focal Mechanisms

The differences in source locations between the mainshock and aftershocks are significant in the EGT inversion. To compensate for this discrepancy and treat each event as a point source at the same location, we horizontally and vertically rotate the focal mechanisms, referring to the literature [4, 5]. Through the horizontal rotation, the station azimuths of the mainshock and aftershocks are set to 90 deg. as measured from north [6]; thus, the number of Green's tensor spatial derivative elements is reduced to three ($G_{i1} = G_{i4} = 0$) for the radial component ($i = y$) and vertical component ($i = z$) and two ($G_{i2} = G_{i3} = G_{i5} = 0$) for the transverse component ($i = x$). Through the vertical rotation, the discrepancy in the takeoff angles between the mainshock and aftershocks is corrected, following the horizontal rotation. The moment tensor elements derived from focal mechanisms should be evaluated after horizontal and vertical rotations. Details of these rotations applied to focal mechanisms should be referred to Ohori and Hisada [6, 8].

8.1.1.4 Correction Applied to the Waveform Data

To adjust the timing between the mainshock and aftershocks, we apply a time-shift to the observation data of aftershocks to fit their S-wave arrival time with that of the mainshock. In addition, to remove the discrepancy in the source time function, we deconvolve the observation data. The observed waveforms used in the estimation of the EGTDs are corrected such that the source time function has a constant seismic moment (1.0×10^{15} Nm, nearly equal to $M_w4.0$) and a single-isosceles slip velocity function with a rise time of 0.32 s. It is noted that the timing between the mainshock and aftershocks and the source time function mentioned above are estimated from 0.2 to 1.0 Hz band-pass-filtered velocity waveforms. We estimated the corner frequencies of events from the records at HYG004, assuming that source spectrum obeys the ω^{-2} model [10]. The corner frequency of the mainshock is about 1.0 Hz, while those of 11 aftershocks are distributed in a frequency range between 1.2 and 3.0 Hz. To simulate the broadband ground motion up to the frequency beyond the corner frequency, we must remove the effect of the differences among corner frequencies. In the present study, to correct the difference in the corner frequency of each event, we assumed the scaling law based on the ω^{-2} model [10] and compensated the spectral amplitude decay beyond the corner frequency of each event so that we can assume that each event has the same corner frequency as that of the mainshock.

8.1.1.5   EGTD Estimation

The observation data for 11 aftershocks are corrected in terms of the timings, source time functions, the seismic moments, and corner frequencies, and they are inverted for the estimation of the EGTD elements. On the basis of the focal mechanisms of aftershocks rotated as mentioned above, simultaneous linear equations for each component are constructed and solved for each of the sampling data. No constraints such as smoothing or minimization for unknown parameters were included in the EGTD estimation in this study. In Fig. 8.2, we show the transverse component elements of the EGTD as example. Each element is scaled for an event with a seismic moment of 1.0×10^{15} Nm with a corner frequency of 1.0 Hz as the same of

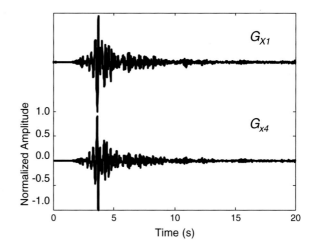

Fig. 8.2 The EGTD elements of transverse components estimated from 11 aftershocks

the mainshock. It is noted that as the Green's tensor spatial derivative elements are determined not by the source characteristics but by the underground structure, the EGTD elements could be useful for the structural study in future work.

8.1.1.6   Simulation of the Strong Ground Motion Using the EGTD

In Fig. 8.3, we compare radial and transverse component observed velocity wave-forms with a 0.2–10 Hz band-pass filtering and corresponding syntheses calculated from the EGTD. For each trace, the source time function, seismic moment, and the corner frequency of each event are taken into account. The top trace for the mainshock (Event 1) is not included in the EGTD inversion. Considering the complexity included in high-frequency components, the broadband synthesis from the EGTD reproduced acceptably the observed waveforms. Figure 8.4 shows the comparison of the maximum amplitude ratio between the synthesized velocity waveforms and observatory data. Three frequency ranges of band-pass filter are compared in Fig. 8.4: 0.2–1.0 Hz, 1.0–10 Hz, and 0.2–10 Hz. From this figure, it is found that simulation results from the EGTD show high accuracy in a frequency range of 0.2–1.0 Hz, except that the radial components of the Event 14 is somewhat overestimated. We also find that results for a frequency range of 1.0–10 Hz are acceptable. For most of the events, the maximum amplitude ratio between the synthesized velocity waveforms and observatory data is between 0.5 and 1.5. We note that results for a frequency range of 0.2–10 Hz are very similar to those of 1.0–10 Hz. On the whole, our simulation of broadband ground motion from the EGTD method reproduced successfully the observed waveforms.

8.2   Conclusions

We demonstrated the applicability of the EGTD method to simulate near-field strong-motion records. In the previous studies [6, 8], considering the lowest corner frequency of about 1.0 Hz for the mainshock of the 2001 Hyogo-ken Hokubu earthquake ($M_j5.4$), we targeted 0.2–1.0 Hz band-pass-filtered velocity waveforms. In present study, we simulate the broadband strong motions between 0.2 and 10 Hz for the same events as in the previous studies [6, 8]. The upper limit of the target frequency range in the EGTD estimation is extended to 10 Hz, while the corner frequency of the events is in a range of 1.0 Hz to 3.0 Hz. So as to correct the discrepancy among the corner frequencies of events, we assumed the scaling law based on the ω^{-2} model [10] and compensated the spectral amplitude decay beyond the corner frequency. The agreement between the observed and calculated wave-forms for the mainshock is satisfactory over a long duration, and there is a good match of the amplitude. To enhance the applicability of the EGTD method, further data accumulation and investigation should be required.

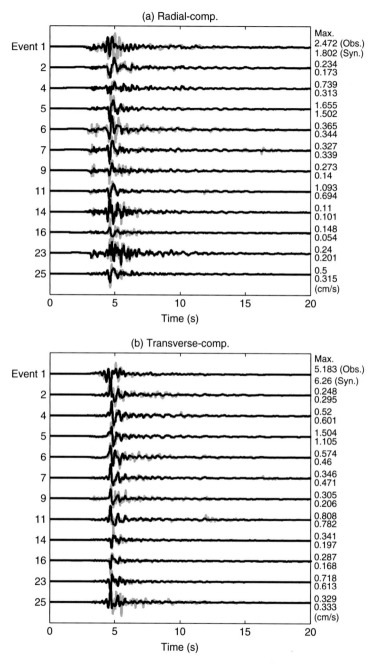

Fig. 8.3 Comparison of 0.2–10 Hz band-pass-filtered observed velocity waveforms (*thick lines*) and corresponding syntheses calculated from the EGTD (*thin lines*)

Fig. 8.4 Comparison of the maximum amplitude ratio between the synthesis and observatory data

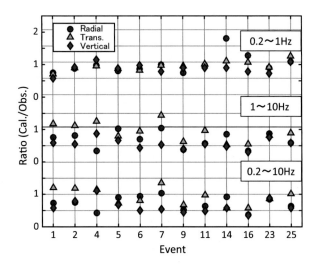

Acknowledgements The strong-motion data used in this study were recorded at the K-NET and the KiK-net stations, provided by the National Research Institute for Earth Science and Disaster Prevention (NIED) on their website (http://www.k-net.bosai.go.jp/k-net/ and http://www.kik.bosai.go.jp/kik/, last accessed December 2005). The Japan Meteorological Agency (JMA) unified hypocenter catalog, and the F-net source parameters were also provided by the NIED on their website (http://www.fnet.bosai.go.jp/, last accessed December 2005). This study was partially supported by Grants-in-Aid for Scientific Research (C) (24540464).

References

1. Hartzell SH (1978) Earthquake aftershocks as Green's functions. Geophys Res Lett 5:1–5
2. Irikura K (1983) Semi-empirical estimation of strong ground motions during large earthquakes. Bull Disaster Prev Res Inst Kyoto Univ 33:63–104
3. Plicka V, Zahradnik J (1998) Inverting seismograms of weak events for empirical Green's tensor derivatives. Geophys J Int 132:471–478
4. Ito Y, Okada T, Matsuzawa T, Umino N, Hasegawa A (2001) Estimation of stress tensor using aftershocks of 15 September 1998 M5.0 Sendai, NE Japan, earthquake. Bull Earthq Res Inst 76:51–59 (in Japanese)
5. Ito Y (2005) A study on focal mechanisms of aftershocks. Rep Natl Res Insti Earth Sci Disaster Prev 68:27–89 (in Japanese)
6. Ohori M, Hisada Y (2006) Estimation of empirical green's tensor spatial derivatives using aftershocks of the 2001 Hyogo-ken Hokubu Earthquake and simulation of mainshock (M_J5.4) strong motion. Zisin 2(59):133–146 (in Japanese with English abstract)
7. Pulido N, Dalguer L, Fujiwara H (2006) Strong motion simulation on a dynamic fault rupture process and empirical Green's tensor derivatives. Fall Meet Seismol Soc Japan, D018

8. Ohori M, Hisada Y (2011) Comparison of the empirical Green's spatial derivative method empirical Green's function method. Bull Seismol Soc Am 101:2872–2886
9. Aki K, Richards PG (1980) Quantitative seismology, theory and methods. W. H. Freeman, San Francisco, 932 pp
10. Aki K (1967) Scaling law of seismic spectrum. J Geophys Res 72:1217–1231

Part III
Probabilistic Risk Assessment with External Hazards

Chapter 9
Development of Risk Assessment Methodology Against External Hazards for Sodium-Cooled Fast Reactors

Hidemasa Yamano, Hiroyuki Nishino, Yasushi Okano, Takahiro Yamamoto, and Takashi Takata

Abstract In this study, hazard evaluation methodologies were developed for the decay heat removal of a typical sodium-cooled fast reactor in Japan against snow, tornado, wind, volcanic eruption, and forest fire. In addition, probabilistic risk assessment and margin assessment methodologies against snow were developed as well. Snow hazard curves were developed based on the Gumbel and Weibull distributions using historical records of the annual maximum values of snow depth and daily snowfall depth. Wind hazard curves were also evaluated using the maximum wind speed and instantaneous speed. The tornado hazard was evaluated by an excess probability for the wind speed based on the Weibull distribution multiplied by an annual probability of the tornado strike at a target plant. The volcanic eruption hazard was evaluated using geological data and tephra diffusion simulation which indicated tephra layer thickness and tephra diameter. The forest fire hazard was evaluated based on numerical simulation which contributed to creating a response surface of frontal fire intensity and Monte Carlo simulation for excess probability calculation. After developing an event tree and failure probabilities, the snow PRA showed the order of 10^{-7}/year of core damage frequency. Event sequence assessment methodology was also developed based on plant dynamics analysis coupled with continuous Markov chain Monte Carlo method in order to apply to the event sequence against snow. Furthermore, this study developed the snow margin assessment methodology that the margin was regarded as the snowfall duration to the decay heat removal failure which was defined as when the snow removal speed was smaller than the snowfall speed.

H. Yamano (✉) • H. Nishino • Y. Okano
Japan Atomic Energy Agency, 4002 Narita-cho, Oarai, Ibaraki 311-1393, Japan
e-mail: yamano.hidemasa@jaea.go.jp

T. Yamamoto
Geological Survey of Japan, AIST, 1-1-1 Higashi, Tsukuba, Ibaraki 305-8567, Japan

T. Takata
Osaka University, 2-1 Yamada-oka, Suita, Osaka 565-0871, Japan

© The Author(s) 2016   111
K. Kamae (ed.), *Earthquakes, Tsunamis and Nuclear Risks*,
DOI 10.1007/978-4-431-55822-4_9

Keywords Probabilistic risk assessment • Sodium-cooled fast reactor • External hazard • Snow • Tornado • Wind • Rain • Volcanic eruption • Forest fire

9.1 Introduction

External hazard risk is increasingly being recognized as important for nuclear power plant safety after the Fukushima Daiichi nuclear power station accident. To improve nuclear plant safety, risk assessment methodologies against various external hazards are necessary, although a probabilistic risk assessment (PRA) methodology against an earthquake has been developed as a priority because of the importance of consequence of an earthquake. The Atomic Energy Society of Japan published a seismic PRA standard in 2007 [1] and a tsunami PRA standard in 2012 [2] which was vigorously developed as an important issue after the Fukushima Daiichi accident caused by a tsunami. Except for the two external hazards, there are no PRA standards against various external hazards in Japan. An alternative methodology different from the PRA was developed in Europe for complementary safety assessments, so-called stress tests [3]. This methodology was useful to show a margin to core damage against earthquake and flood. Since the most challenge in developing external PRA methodologies is to quantify the intensity of the external hazards for the assessment, the stress test methodology would be useful and effective to suggest safety measures and accident managements that can extend margins to core damage against external hazards.

This study aims mainly at a contribution to the risk assessment and safety improvement of the decay heat removal function of a prototype sodium-cooled fast reactor (SFR) in Japan. It is well known that an earthquake is the most important external hazard that would have a potential structural impact on system, structure, and components of plants. In typical light water reactors (LWRs), flooding including tsunami is an important hazard because its heat sink is sea (river), which is also well known after the Fukushima Daiichi nuclear power station accident. On the other hand, the external flooding is not so significant in SFRs of which heat sink is air. The decay heat removal system (DHRS) of the SFR utilizes air coolers (ACs) located at high elevation, which might be affected by aboveground hazards. This study also takes into account effects on ventilation and air-conditioning system, emergency power supply system, and so on, for which air is usually taken.

This study addresses extreme weathers (snow, tornado, wind, and rainfall), volcanic phenomena, and forest fire as representative aboveground external hazards, which was selected through a screening process [4]. In the first screening, after all foreseeable external hazards were exhaustively identified, a wide variety of external hazards were screened out in terms of site conditions, impact on plant, progression speed, envelop, and frequency. In the second screening, the external hazards were selected on a basis of the scope of this study (aboveground natural hazards). Similar hazards were also merged; e.g., hail can be enveloped by tornado-induced missiles. Combination of external hazards is very important in the risk assessment. For instance, in terms of aboveground hazards, this study would address the following hazards: strong wind and heavy rain, snow and cold temperature, volcanic eruption and rain, and so on.

When an extreme external hazard occurs, the nuclear plant is expected to be shut down normally. Therefore, only the decay heat removal function was taken into account, assuming success of reactor shutdown in this study. Although the Fukushima Daiichi accident lessons suggested the importance of a spent fuel pool, this study focuses as a first step on event sequences resulting in reactor core damage because a grace period of accident management is short under hot condition in a full-power operation. The developed methodology is applied mainly to SFRs, though it would also be basically applicable for LWRs in which air is necessary for emergency diesel generators.

The objective of this study is to develop both the margin assessment and PRA methodologies against the representative external hazards. The overview of this study is schematically illustrated in Fig. 9.1. The PRA would indicate a core damage frequency (CDF), which calculates a summation of conditional heat removal failure probabilities multiplied by hazard occurrence frequencies which is based on a hazard curve representing relation between the frequency and the hazard intensity. The margin assessment would show the extension of margins from a design basis to the core damage by introducing several measures including accident management. An advantage of the margin assessment methodology is un-necessity of quantitative external hazard evaluation. Since the event sequence evaluation is needed both for the margin assessment and PRA, a difference between them is quantification of external hazards.

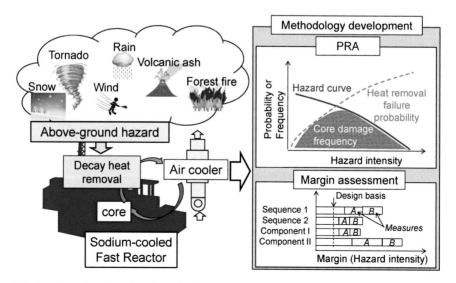

Fig. 9.1 Overview of methodology development

The present paper is intended to develop hazard evaluation methodologies against snow, tornado, wind, volcanic eruption, and forest fire. This paper also describes the PRA and margin assessment methodologies against snow.

9.2 External Hazard Evaluation Methodologies

9.2.1 Snow

The snow hazard indexes are the annual maximum snow depth and the annual maximum daily snowfall depth. Snow hazard curves for the two indexes were developed using 50-year historical weather records at the prototype SFR site which is located in Japan Sea side central area [4].

In this study, a snow hazard evaluation methodology was developed according to the following procedure. At first, the annual maximum data of the snow depth and daily snowfall depth were collected. Using these data, the annual excess probability was evaluated by plotting position formula Weibull, Hazen, and Cunnane for general use. Of the three formulas, it is said that the Cunnane is the best suitable and applicable to all probability distributions. Next, the parameters of Gumbel or Weibull cumulative probability distributions were determined by a least square method. Using the annual excess probability, the snow hazard curves were successfully obtained after checking the conformance and stability evaluations in terms of the annual maximum snow depth and the annual maximum daily snowfall depth. Figure 9.2 shows the snow hazard curves using the Gumbel and Weibull distributions. It should be noted that the difference between the two distributions becomes large in a low-frequency range exceeding the measured data ($\sim 10^{-2}$/year). This may be caused by epistemic uncertainty (i.e., lack of knowledge). Considering this uncertainty, conservative evaluations or sensitivity analysis is useful and recommended in the risk assessment.

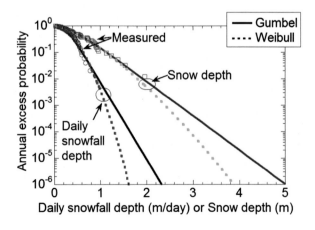

Fig. 9.2 Snow hazard curve

9.2.2 Tornado

Wind scales estimated based on structural damage caused by a tornado are usually represented by Fujita scale, which is defined as F0 = 17–32 m/s (average time: ~15 s), F1 = 33–49 m/s (~10s), F2 = 50–69 m/s (~7 s), F3 = 70–92 m/s (~5 s), F4 = 93–116 m/s (~4 s), and F5 = 117–142 m/s (~3 s).

The procedure of estimation for the tornado hazard curve is as follows. The first step is to select an area for estimating a tornado hazard curve in Japan and to analyze historical tornado data recorded in the selected area. This study selects one of the areas along the seashore of Japan Sea including Hokkaido, which includes the SFR site. The range to collect tornado data is 5 km inland and sea from the seashore in this area. The second step is to estimate the annual probability of the tornado strike at the target nuclear plant. The third step is to estimate the excess probability for maximal wind speed calculated from Weibull distribution. The final step is to multiply the annual probability estimated in the second step by the excess probability estimated in the third step. By this calculation, the tornado hazard curve was successfully estimated [5].

9.2.3 Strong Wind

The wind hazard index is the annual maximum instantaneous wind speed which is used to estimate missile speed [6]. Likewise the snow hazard curve, a basic concept of this methodology is a generalized estimation way, which is characterized by obtaining appropriate probability distribution through the conformance and stability evaluations.

After the collection of wind speed data, an annual excess probability distribution can be evaluated by using wind data based on plotting position formula. The strong wind hazard curves were developed using the Gumbel and Weibull distributions, of which parameters were calculated by a least square method. Figure 9.3 shows the hazard curves based on the Gumbel and Weibull distributions. In the Gumbel distribution, the estimated curve decreases linearly less than 0.1 of the annual excess probability. In the Weibull distribution, on the other hand, the curve decreases like a quadratic curve. From this figure, the larger the difference between the two estimated distributions is, the lower the excess probability is.

9.2.4 Volcanic Eruption

Volcanic ash was identified as the key phenomena of the volcanic eruption hazard in the vicinity of the plant site, so that the volcanic ash hazard evaluation methodology is being developed using geological data and numerical simulations of ash diffusion. Geological data survey indicated about 2×10^{-4}/year of volcanic ash

116   H. Yamano et al.

Fig. 9.3 Wind hazard
curve

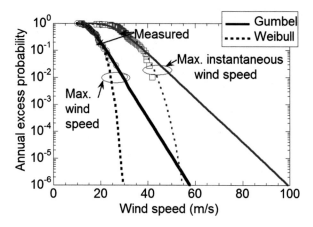

Fig. 9.4 Calculated tephra
layer thickness and tephra
diameter

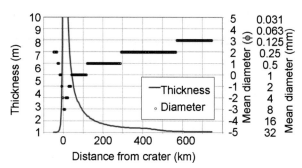

fallout around the site, which was based on boring data of 22 ash fallouts since
110,000 years ago. This includes thin ash layers. For thicker ash layers than 5 cm,
the volcanic ash fallout frequency was estimated about 2×10^{-5}/year.

From the geological data, the maximum thickness of the ash fallouts around the
plant site is about 50 cm of the Daisen-Kurayoshi tephra that erupted about
50,000 years ago. This study carried out numerical simulation of the Daisen-Kurayoshi
tephra diffusion using the Tephra2 code [7]. The simulation showed well-reproduced
ash fallout distribution in a wide area. Figure 9.4 shows calculated fallout thickness and
tephra diameter along the distance from the crater. A discharge rate of fallout was
estimated about 10^{19} kg/s, and eruption duration was 4–8×10^{4} s [8].

One of crucial issues of volcanic eruption is to investigate characterization of
volcanic particle, in particular fine volcanic ash less than 0.06 mm in diameter,
which could disperse vast area from the source volcano and be easily remobilized
by surface wind and precipitation after the deposition. In order to quantify quantitative characteristics of fine volcanic ash particle, we sampled volcanic ash directly
falling from the eruption plume from Sakurajima volcano before landing on ground.
A newly introduced high precision digital microscope and particle grain size
analyzer allowed us to develop hazard evaluation method of fine volcanic ash
particle [9].

9.2.5 Forest Fire

The phenomena of a forest fire that would give potential impacts on nuclear plants are identified as fire, flame, smoke, and flying objects. To evaluate their impacts, numerical simulations are utilized by using the FARSITE code with appropriate numerical conditions: fire breakout, fire spread condition including extinguishing, weather data, vegetation data, and simulation conditions. For these conditions, branch probabilities are provided based on a logic tree. The simulation showed that the wind speed and relative humidity were sensitive to the forest fire hazard [10].

A preliminary hazard evaluation was carried out using a response surface of frontal fire intensity with regard to the wind speed and relative humidity. The evaluated hazard curve is such that the annual excess probability is about 1.0×10^{-4}/year for the frontal fireline intensity of 200 kW/m and about 1.3×10^{-5}/year for 300 kW/m [11].

Smoke is also important in the forest fire hazard evaluation. The ALOFT-FT code was applied to the smoke transport analysis in order to investigate potential impact on air filters for the DHRS. The total amount of particle matters estimated was estimated well below the operational limit of the air filter [12].

9.3 Risk Assessment Methodologies Against Snow

9.3.1 PRA

Snow hazard categories were obtained from a combination of the daily snowfall depth (snowfall speed) and snowfall duration that can be calculated by dividing the snow depth by the snowfall speed [13]. For each snow hazard category, accident sequences were evaluated by producing event trees that consist of several headings representing the loss of the decay heat removal. Air ventilation channels must be ensured for the important components in this PRA: emergency diesel generator, ACs in the decay heat removal system. The natural circulation decay heat removal is expected in the SFR, so that manual operation of the AC dampers is required in a total blackout situation (the loss of direct current-powered equipment). Snow removal operation was introduced into the event trees as the accident managements. To succeed in the snow removal, plant personnel have to be able to reach the door to open on the building roof and then have to remove accumulated snow from the door to the air inlets. The failure probabilities were evaluated as a function of hazard intensity.

The decay heat removal failure probability of each event sequence was obtained by introducing the failure probability into the event tree. The CDF by the snow hazard category can be calculated by multiplying each heat removal failure probability and each snow hazard occurrence frequency. In total, the CDF brings the

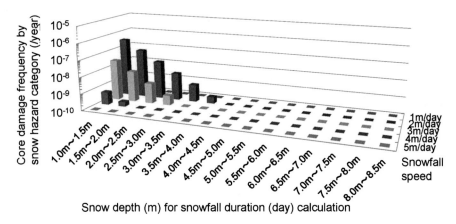

Fig. 9.5 Core damage frequency by snow hazard category

order of 10^{-7}/year. Figure 9.5 shows the CDF by the snow hazard category, in which the dominant snow hazard category was a combination of 1–2 m/day of snowfall speed and 0.5–0.75 day of snowfall duration (1–1.5 m of snow depth). The dominant sequence was that the personnel failed the door opening on the roof after the 1st awareness of the snow removal necessity, resulting in the loss of decay heat removal system due to snow. Importance and sensitivity analyses indicated a high risk contribution to secure the access routes.

Looking at Fig. 9.2, the dominant snowfall speed of 1–2 m/day is approximately 10^{-2}/year of annual access probability (at 1 m/day of snowfall), and the dominant snow depth of 1–1.5 m is approximately 10^{-1}/year (at 1 m of snow depth). Such frequencies are not so low that we are aware of the importance of relatively high frequent hazard through this study. The PRA results would be served for the development of safety measures and accident management. In general, although careful attention may be often paid to extremely low-frequency events bringing high consequence, significant hazard intensity could be clarified through PRA studies.

The event tree methodology is well known as a classical manner for the PRA; however, it is difficult to express time-dependent event sequences including recovery. Therefore, a new assessment technique was also being developed for the event sequence evaluation based on a continuous Markov chain Monte Carlo method with plant dynamics analysis [14].

9.3.2 Margin Assessment

We introduced an effective snow removal speed which is defined as a daily snow removal speed multiplied by a performance factor of the snow removal work so that plant personnel can remove accumulated snow in a certain time. If this effective

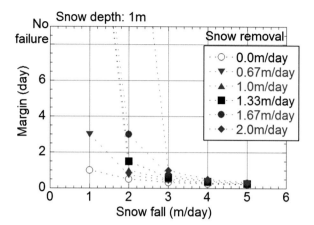

Fig. 9.6 Margin assessment against snow

snow removal speed exceeds the snowfall speed, the scenario leads to no heat removal failure. On the other hand, if the effective snow removal speed falls below the snowfall speed, the heat removal failure scenario appears as a result of gradual continuous accumulation of snow. The margin (day) can be defined as the snowfall duration until when the accumulated snow depth reaches the snow depth corresponding to the heat removal failure. In this definition, the accumulated snow depth can be calculated as a difference between the snowfall speed and the effective snow removal speed.

This study assumed the failure to secure the access routes when the snow depth reached 1 m as well as conservatively assumed 1 m for the snow depth corresponding to the heat removal failure in the decay heat removal system. The performance factor of the snow removal work was set 1/3, assuming totally 8 h per day for plant personnel to remove the snow. These assumptions were applied to the margin assessment. The margin assessment result is presented in Fig. 9.6, in which the parameter is the effective snow removal speed. No heat removal failure appears if the snow removal speed is higher than 3 m/day (1 m/day of effective snow removal speed) when the snowfall speed is 1 m/day. Even if the same snow removal speed is applied, the heat removal failure scenario appears to indicate 1 day of margin when the snowfall speed is 2 m/day. Considering such a situation, it is important to flexibly strengthen a snow removal action plan such as an increase in the performance factor of the snow removal work.

9.4 Conclusion

In this study, hazard evaluation methodologies were developed for the decay heat removal function of a typical sodium-cooled fast reactor in Japan against snow, tornado, wind, volcanic eruption, and forest fire. In addition, PRA and margin assessment methodologies against snow were developed as well. Snow hazard curves were developed based on the Gumbel and Weibull distributions using historical records of the annual maximum values of snow depth and daily snowfall depth. Likewise this snow hazard evaluation, wind hazard curves were also developed using the maximum wind speed and instantaneous speed. The tornado hazard data were used based on the Fujita scale in a wide area along the seashore of Japan Sea. Using the historical records, the tornado hazard curve was estimated as an excess probability for the wind speed based on the Weibull distribution multiplied by an annual probability of the tornado strike at the target nuclear plant. The volcanic eruption hazard was evaluated using geological data and tephra diffusion simulation which indicated tephra layer thickness and tephra diameter. The forest fire hazard was evaluated based on numerical simulation which contributed to creating a response surface of frontal fire intensity with regard to the wind speed and relative humidity, etc., and Monte Carlo simulation for excess probability calculation. For the snow PRA, the accident sequence was evaluated by producing event trees which consist of several headings representing the loss of decay heat removal. Snow removal action and manual operation of the air cooler dampers were introduced into the event tree as accident managements. In this paper, the snow PRA showed the order of 10^{-7}/year of core damage frequency. The dominant snow hazard category was the combination of 1–2 m/day of snowfall speed and 0.5–0.75 day of snowfall duration. Event sequence assessment methodology was also developed based on plant dynamics analysis coupled with continuous Markov chain Monte Carlo method in order to apply to the event sequence against snow. For the snow margin assessment, the index is the combination of a snowfall speed and duration. Since snow removal can be expected for the snowfall, this study developed the snow margin assessment methodology that the margin was regarded as the snowfall duration until when the accumulated snow depth reaches the snow depth corresponding to the decay heat removal failure.

Acknowledgments The present study is the result of "Research and Development of Margin Assessment Methodology of Decay Heat Removal Function against External Hazards" entrusted to Japan Atomic Energy Agency by the Ministry of Education, Sports, Science and Technology (MEXT) in Japan.

References

1. Atomic Energy Society of Japan (AESJ) (2007) Implementation standard of seismic probabilistic safety assessment for nuclear power plants: 2007. AESJ-SC-P006:2007
2. Atomic Energy Society of Japan (AESJ) (2011) Implementation standard concerning the tsunami probabilistic risk assessment of nuclear power plants: 2011. AESJ-SC-RK004E:2011

3. European Nuclear Safety Regulators Group (ENSREG) (2012) Peer review report; stress tests performed on European nuclear power plants. 25 Feb 2012 (online)
4. Yamano H, Nishino H, Kurisaka K, Sakai T, Yamamoto T, Ishizuka Y, Geshi N, Furukawa R, Nanayama F, Takata T (2014) Development of margin assessment methodology of decay heat removal function against external hazards – project overview and preliminary risk assessment against snow. In: Proceedings of the 12th probabilistic safety assessment and management conference (PSAM 12), Honolulu, Hawaii, USA, 22–27 June 2014, No. 44
5. Nishino H, Kurisaka K, Yamano H (2014) Development of margin assessment methodology of decay heat removal function against external hazards (2) Tornado PRA methodology. In: Proceedings of the 10th international topical meeting on nuclear thermal-hydraulics, operation and safety (NUTHOS-10), Okinawa, Japan, 14–18 Dec 2014, NUTHOS10-1068
6. Yamano H, Nishino H, Kurisaka K, Okano Y, Sakai T, Yamamoto T, Ishizuka Y, Geshi N, Furukawa R, Nanayama F, Takata T, Azuma E (2015) Development of risk assessment methodology against external hazards for sodium-cooled fast reactors: project overview and strong wind PRA methodology. In: Proceedings of the international congress on advances in nuclear power plants (ICAPP2015), Nice, France, 3–6 May 2015, No. 15031
7. Connor CB, Hill BE, Winfrey B, Franklin NM, Lafemina PC (2001) Estimation of volcanic hazards from tephra fallout. Nat Hazards Rev 2:33–42
8. Yamamoto T, Sugiyama M, Tajima Y (2014) Simulation of the Daisen-Kurayoshi Tephra, in the San-in District, SW Japan, using Tephra2. Japan Geoscience Union Meeting 2014, 28 Apr– 2 May 2014, SVC52-01
9. Nanayama F, Furukawa R, Ishizuka Y, Yamamoto T, Geshi N, Oishi M (2013) Characterization of fine volcanic ash from explosive eruption from Sakurajima Volcano, South Japan. Transaction of American Geophysical Union 2013 Fall Meeting, San Francisco, CA, USA, 9–13 Dec 2013
10. Okano Y, Yamano H (2014) Development of margin assessment methodology of decay heat removal function against external hazards (3) Forest fire hazard assessment methodology. In: Proceedings of the 10th international topical meeting on nuclear thermal-hydraulics, operation and safety (NUTHOS-10), Okinawa, Japan, 14–18 Dec 2014, NUTHOS10-1018
11. Okano Y, Yamano H (2015) Development of a hazard curve evaluation method for a forest fire as an external hazard. In: Proceedings of the international topical meeting on probabilistic safety assessment and analysis (PSA 2015), Sun Valley, ID, USA, 26–30 Apr 2015
12. Okano Y, Yamano H (2015) Development of risk assessment methodology of decay heat removal function against external hazards for sodium-cooled fast reactors (3) Numerical simulations of forest fire spread and smoke transport as an external hazard assessment methodology development. In: Proceedings of the 23rd international conference on nuclear engineering (ICONE-23), Chiba, Japan, 17–21 May 2015, ICONE23-1009
13. Yamano H, Nishino H, Kurisaka K, Okano Y, Sakai T, Yamamoto T, Ishizuka Y, Geshi N, Furukawa R, Nanayama F, Takata T, Azuma E (2014) Development of margin assessment methodology of decay heat removal function against external hazards (1) Project overview snow PRA methodology. In: Proceedings of the 10th international topical meeting on nuclear thermal-hydraulics, operation and safety (NUTHOS-10), Okinawa, Japan, 14–18 Dec 2014, NUTHOS10-1014
14. Takata T, Azuma E (2014) Development of margin assessment methodology of decay heat removal function against external hazards (4) Event sequence assessment based on continuous Markov Chain Monte Carlo method with plant dynamics analysis. In: Proceedings of the 10th international topical meeting on nuclear thermal-hydraulics, operation and safety (NUTHOS-10), Okinawa, Japan, 14–18 Dec 2014, NUTHOS10-1291

Chapter 10
Effectiveness Evaluation About the Tsunami Measures Taken at Kashiwazaki-Kariwa NPS

Masato Mizokami, Takashi Uemura, Yoshihiro Oyama, Yasunori Yamanaka, and Shinichi Kawamura

Abstract All of the nuclear power stations of TEPCO had experienced huge external events. One of which is the Niigata-ken Chuetsu-Oki earthquake in 2007 at Kashiwazaki-Kariwa Nuclear Power Station (NPS), and the other is the Great East Japan Earthquake in 2011 at Fukushima Daiichi NPS and Fukushima Daini NPS. Especially, the Fukushima Daiichi Units 1–3 experienced severe accident, since prolonged station blackout (SBO) and loss of ultimate heat sink (LUHS) were induced by the huge tsunami which was generated by the Great East Japan Earthquake. The most important lesson learned was that the defense-in-depth for external event was insufficient. Therefore, we are implementing many safety enhancement measures for tsunami in our Kashiwazaki-Kariwa Nuclear Power Station. Thus, in order to confirm the effectiveness of these safety enhancement measures, TEPCO performed tsunami PRA studies. The studies were conducted in accordance with "The Standard of Tsunami Probabilistic Risk Assessment (PRA) for nuclear power plants" [1] established by the Atomic Energy Society of Japan. TEPCO conducted two state (the state before the implementation of accident management (AM) measures and the state at the present) evaluations to confirm the effectiveness of the safety enhancement measures. In this evaluation, TEPCO were able to confirm the effectiveness of safety enhancement measures carried out towards plant vulnerabilities that were found before these measures were implemented.

Keywords Tsunami PRA • Probabilistic risk assessment • Fragility • External event • Kashiwazaki-Kariwa • Fukushima Daiichi accident

M. Mizokami (✉) • T. Uemura • Y. Oyama • Y. Yamanaka • S. Kawamura
Nuclear Asset Mangement Department, Tokyo Electric Power Company, 1-3 Uchisaiwai-cho 1chome Chiyoda-ku, Tokyo, Japan
e-mail: mizokami.masato@tepco.co.jp

K. Kamae (ed.), *Earthquakes, Tsunamis and Nuclear Risks*,
DOI 10.1007/978-4-431-55822-4_10

123

10.1   Introduction

All of the nuclear power stations of TEPCO had experienced huge external events. One of which is the Niigata-ken Chuetsu-Oki earthquake in 2007 at Kashiwazaki-Kariwa Nuclear Power Station (NPS), and the other is the Great East Japan Earthquake in 2011 at Fukushima Daiichi NPS and Fukushima Daini NPS. Especially, the Fukushima Daiichi Units 1–3 experienced severe accident, since prolonged station blackout (SBO) and loss of ultimate heat sink (LUHS) were induced by the huge tsunami which was generated by the Great East Japan Earthquake. One of the lessons learned is "defense-in-depth for tsunami was insufficient." In terms of safety enhancement of nuclear power plant from this lesson, countermeasure for each layer of defense-in-depth against tsunami is enhanced in the Kashiwazaki-Kariwa NPS. Then, we perform tsunami PRA in order to understand plant vulnerability and to check validity of deployed countermeasure against tsunami for Unit 7 (ABWR) of the Kashiwazaki-Kariwa NPS. This paper describes the evaluation result completed by applying to states before and after the implementation of the tsunami countermeasures.

10.2   Outline of Kashiwazaki-Kariwa Nuclear Power Station

The Kashiwazaki-Kariwa Nuclear Power Plant (see Fig. 10.1) is located in Kariwa Village and Kashiwazaki City in Niigata Prefecture facing on the coast of the Japan Sea, and seven nuclear reactors (Unit 1–5: BWR5, Unit 6, 7: ABWR, a total of 8212 MWe) are built.

The ground elevation is T.P. 5 m (Tokyo Peil: sea-level of Tokyo Bay) at the north side (Units 1–5) and T.P. 12 m at the south side (Units 5–7).

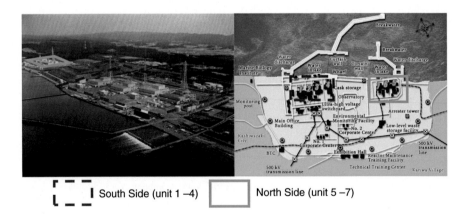

South Side (unit 1 –4)   North Side (unit 5 –7)

Fig. 10.1 Kashiwazaki-Kariwa NPS

10.3 Tsunami PRA for Kashiwazaki-Kariwa Nuclear Power Station

In Japan, from the lesson of the Fukushima Daiichi accident, development of tsunami PRA method was accelerated immediately after the accident, and Atomic Energy Society of Japan (AESJ) issued tsunami PRA guideline in February 2012 [1]. Then, TEPCO started to perform tsunami PRA to evaluate the effectiveness of tsunami countermeasures. In the state before the implementation of tsunami countermeasures, since there is no means to prevent flooding to building and function failure of important equipment assuming generation of tsunami exceeding the 1st floor height of the building, each flooding propagation evaluation and fragility evaluation is done with a simple method, and the core damage frequency (CDF) for each accident sequence is calculated.

10.3.1 Tsunami Hazard Evaluation

Tsunami hazard for the Kashiwazaki-Kariwa NPS is evaluated based on the "method of probabilistic tsunami hazard analysis" [2] issued in 2009 by the Japan Society of Civil Engineers (JSCE). However, the occurrence frequency and the scale of earthquake, assuming multi-segment rupture of the faults which is the latest knowledge acquired in the 2011 off the Pacific coast of Tohoku Earthquake, are also taken into consideration.

10.3.1.1 Tsunami Source Model

Regarding the tsunami-induced source area, the tsunami induced by earthquake, originated by faults which exist in the area, is determined in terms of whether they have significant influence on the tsunami hazard of the Kashiwazaki-Kariwa NPS. As a result, the following areas are selected:

1. The fault which is considered in seismic design and is identified by geological survey, etc.
2. The fault which is unidentified by investigation, but indicated by an external organization (epicenter at coast of the Niigata southwest earthquake).
3. The east edge of Japan Sea; Kashiwazaki-Kariwa NPS is considered to be affected significantly when tsunami occurs there.

Regarding these tsunami occurrence areas, the tsunami occurrence scenario is created by setting up the magnitude range and the earthquake recurrence interval.

10.3.1.2 Uncertainty

Random uncertainty in a numerical computation model and epistemic uncertainty regarding some issues such as the existence of active fault and magnitude range, etc., are considered in tsunami hazard evaluation. Epistemic uncertainty is dealt with as number of branch of tsunami occurrence scenario, and given weighting to each scenario. Weights of discrete branches that represent alternative hypotheses and interpretations were determined by the JSCE guideline basically. In this evaluation, the magnitude range, earthquake occurrence probability, probability of multi-segment rupture of the faults, and probability distributions of random uncertainty are taken into consideration.

10.3.1.3 Hazard Curve

The annual probability of exceedance of tsunami wave height is created for each tsunami occurrence scenario defined in Sects. 11.3.1.1 and 11.3.1.2. Next, for each curve, with consideration for the weighting corresponding to each scenario, statistical processing is performed, and hazard curve is created for weighted average as arithmetic average for weighted accumulation sum as fractal curve. As mentioned above, the tsunami hazard curve (tsunami run-up area at the north side) is shown in Fig. 10.2. In evaluation of the state before the implementation of tsunami countermeasures, when tsunami exceeds height of the 1st floor of building, it is simply assumed that flooding in the building occurs and equipment function is lost, and it causes core damage. For example, in the evaluation of Unit 7, since the 1st floor

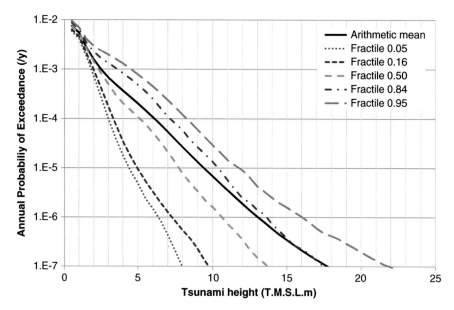

Fig. 10.2 Tsunami hazard curve

height is T.P.12.3 m, when the tsunami beyond this height strikes, it is evaluated as core damage occurs.

10.3.2 Tsunami Fragility Evaluation

Regarding influence of tsunami on equipment, damage by flooding and by tsunami wave force is considered. Regarding equipment on yard and door on outer wall of the buildings such as yard tank, yard watertight door, etc., the failure probability against tsunami wave force is set by flooding depth based on tsunami run-up analysis result. Regarding equipment and door inside building, the damage probability is set by flooding propagation analysis result for building. Regarding tsunami run-up analysis, it is performed for multi-case of tsunami height. For each case, fragility curve is evaluated from the equipment damage probability with consideration for the uncertainty in the flooding depth of the installation location for each equipment. The main assumptions in the fragility evaluation are shown below:

1. Embankment, tidal wall
   When tsunami exceeds the height of the embankment or tidal wall, these failures are assumed.
2. Watertight door, general door
   Regarding protection doors installed on building outer wall, fragility evaluation is conservatively performed with consideration for tsunami wave force.
3. Yard tanks (light oil tank, pure water storage tank)
   Since these tanks are on the ground, damage evaluation by tsunami wave force is performed, but evaluation for flooding and function affected by water level by submersion is also performed.
4. Fire protection system piping
   Fracture evaluation is performed for bending load of piping changed by tsunami wave force. Branch piping which has high failure possibility is also taken into consideration.
5. Equipment in building (reactor core isolation cooling system (RCIC), power panel, etc.)
   Flooding propagation evaluation in building is performed, and when the concerned equipment and required support system are inundated, the function failures are assumed.

However, in evaluation of the state before the implementation of tsunami countermeasures, fragility evaluation with consideration for uncertainty is not performed, but method that the events induced by the tsunami of a certain height are deterministically evaluated is adopted.

10.3.3   Accident Scenario Identification

10.3.3.1   The State Before the Implementation of Tsunami Countermeasures

At the state before the implementation of tsunami countermeasures, it is assumed accident scenarios considering flooding according to the tsunami wave height. In addition, if the tsunami height is below the site level (T.P. 12 m), it is assumed that inundation starts via maintenance hatch (T.P. 3.5 m) in the heat exchanger area in the turbine building when tsunami height exceeds T.P. 3.5 m. Also, it is conservatively assumed that all the buildings connected to turbine building are flooded to the tsunami height.

0. Tsunami height between T.P. 4.2 m and T.P. 4.8 m
   The support system (e.g., reactor cooling water system (RCW) pumps, reactor sea water system (RSW) pumps) is located in basement 1st floor of turbine building (T/B). When tsunami height exceeds T.P. 4.2 m, the support system is flooded, and it causes LUHS by the function failure. In addition, non-safety-related metal-clad switch gear (M/C) in basement 2nd floor of T/B is also flooded.
1. Tsunami height between T.P. 4.8 m and T.P. 6.5 m
   Emergency M/C in basement floor of reactor building (R/B) is flooded and lost its function. It causes SBO by the function failure of emergency M/C and non-safety-related M/C, because it cannot be powered by off-site power and emergency diesel generators (D/Gs).
2. Tsunami height between T.P. 6.5 m and T.P. 12.3 m
   DC power panel in the basement floor of control building (C/B) is flooded and loses its function. It causes loss of DC power.
3. Tsunami height exceeding T.P. 12.3 m
   Tsunami runs up to the site level, low-voltage start-up transformer located at the site level is flooded and loses its function, and inundation into the main buildings occurs via entrance of each building.

10.3.3.2   The State After the Implementation of Tsunami Measures

Using the results of tsunami fragility analysis as a reference, initiating events which are induced by tsunami are adopted and accident scenario analysis is conducted.
   The extracted initiating events are shown below:

1. Loss of off-site power (LOOP)

   • Flooding of low-voltage start-up transformer

2. Loss of function of emergency D/G

- Flooding of emergency D/G(A,B,C) by inundation of R/B
- Fuel transport failure by damage of light oil tank
- Fuel transport failure by damage of fuel transport pump
- Operation failure of emergency D/Gs operation failure by loss of support system function by T/B flooding
- Flooding of emergency power panel room in R/B

3. Loss of ultimate heat sink

- Loss of support system function by T/B flooding
- Loss of support system function by D/G failure (in case of LOOP)

4. Loss of instrumentation and control system function

- Flooding of main control room (MCR) in C/B
- Flooding of DC power panel in C/B

Plant walkdown in R/B, T/B, and yard is implemented by analysts and designers to confirm the result of fragility analysis and assumed accident scenario. As a result, validity of the fragility and scenario is checked.

10.3.4 Accident Sequence Evaluation

10.3.4.1 The State Before the Implementation of Tsunami Countermeasures

Accident scenario changes according to tsunami height. So, initiating events and credited mitigation systems are changed as well.

1. Tsunami height between T.P. 4.2 m and T.P. 4.8 m
   Initiating event is set as LUHS. In identified accident scenario, the relief valve function of SRV and RCIC are credited as mitigation systems. Event tree is shown in Fig. 10.3. CDF for this tsunami height is calculated as 8.8E-5(/RY), and dominant sequence is TQUV (transient with loss of all ECCS injections).
2. Tsunami height between T.P. 4.8 m and T.P. 6.5 m
   Initiating event is set as LUHS and SBO. Credited mitigation system is the same as (1). Event tree is shown in Fig. 10.4. CDF for this tsunami height is calculated as 1.0E-4(/RY) and dominant sequence is TQUV.
3. Tsunami height exceeding T.P. 6.5 m
   Initiating event is set as LUHS, SBO, and loss of DC power. No credited mitigation system is set because it is assumed loss of DC power. Event tree is shown in Fig. 10.5. CDF for this tsunami height is calculated as 2.5E-5 (/RY), and dominant sequence is TBD (transient with loss of all AC and DC powers).

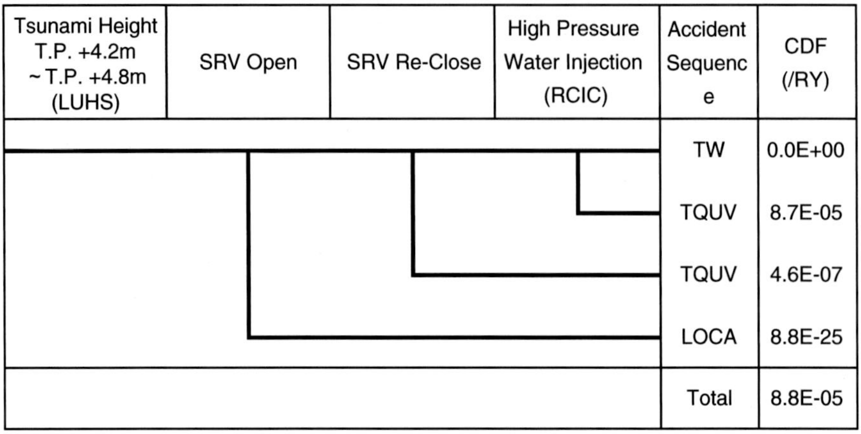

Tsunami Height T.P. +4.2m ~T.P. +4.8m (LUHS)	SRV Open	SRV Re-Close	High Pressure Water Injection (RCIC)	Accident Sequence	CDF (/RY)
				TW	0.0E+00
				TQUV	8.7E-05
				TQUV	4.6E-07
				LOCA	8.8E-25
				Total	8.8E-05

TW: Transient with loss of decay heat removal
    → Containment vessel failure before core damage
TQUV: Transient with loss of all high pressure and low pressure ECCS injections.
LOCA: Loss of Coolant Accident

Fig. 10.3 Event tree (tsunami height T.P. 4.2 m–4.8 m)

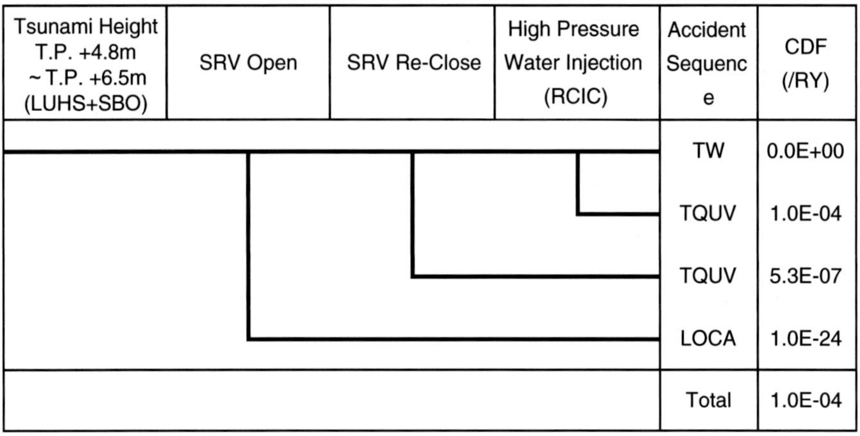

Tsunami Height T.P. +4.8m ~T.P. +6.5m (LUHS+SBO)	SRV Open	SRV Re-Close	High Pressure Water Injection (RCIC)	Accident Sequence	CDF (/RY)
				TW	0.0E+00
				TQUV	1.0E-04
				TQUV	5.3E-07
				LOCA	1.0E-24
				Total	1.0E-04

TW: Transient with loss of decay heat removal
    → Containment vessel failure before core damage
TQUV: Transient with loss of all high pressure and low pressure ECCS injections.
LOCA: Loss of Coolant Accident

Fig. 10.4 Event tree (tsunami height T.P. 4.8 m–6.5 m)

Tsunami PRA results at the state before the implementation of countermeasures is shown in Fig. 10.6. Total CDF is calculated as 2.1E-4(/RY) in average value. As for accident sequence rate, TQUV is dominant sequence accounting for 89 percentages.

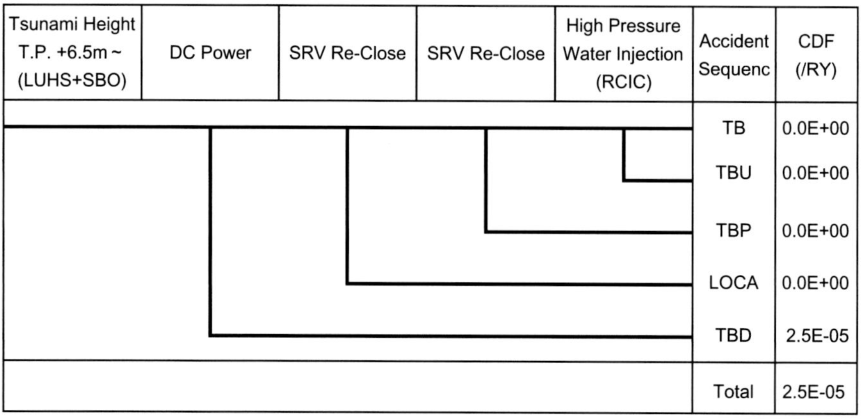

Tsunami Height T.P. +6.5m ~ (LUHS+SBO)	DC Power	SRV Re-Close	SRV Re-Close	High Pressure Water Injection (RCIC)	Accident Sequenc	CDF (/RY)
					TB	0.0E+00
					TBU	0.0E+00
					TBP	0.0E+00
					LOCA	0.0E+00
					TBD	2.5E-05
					Total	2.5E-05

TB: Transient with loss of all AC powers → depletion of DC powers and loss of ECCS(RCIC) injections
TBU: Transient with loss of all AC powers and ECCS(RCIC) injections
TBP: Transient with loss of all AC powers and failure to SRV closing (no low pressure ECCS injections)
LOCA: Loss of Coolant Accident
TBD: Transient with loss of all AC & DC powers

Fig. 10.5 Event tree (tsunami height exceeding T.P. 6.5 m)

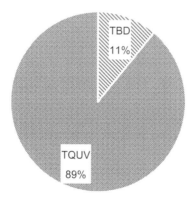

TQUV: Transient with loss of all high pressure and low pressure ECCS injections
TBD: Transient with loss of all AC & DC powers

Fig. 10.6 Contribution of each accident sequences for CDF in tsunami PRA (the state before the implementation of tsunami countermeasures)

10.3.4.2   The State After the Implementation of Tsunami Countermeasures

Based on the result of tsunami fragility analysis, in the accident sequence analysis, failure rate which is relevant to initiating events or equipment relevant to credited mitigation system is calculated, and combination of tsunami height and damaged equipment is considered.

Regarding the accident sequence analysis, tsunami initiating hierarchy event tree is constructed. In this event tree, yard equipment whose failure is directly connected to the initiating event is set as heading. The hierarchy event tree is shown in Fig. 10.7. In event tree for each initiating event which is expanded from the hierarchy event tree, yard equipment which is not considered as heading is set as mitigation systems.

The outline of accident sequence analysis is described below:

1. Tsunami height between T.P. 15 m and T.P. 17 m
   Because, as shown by the fragility analysis result, the watertight doors of each building are not broken by tsunami of this height, inundation into the buildings does not occur, but the fuel transport pumps on yard are destroyed by tsunami. In this state, random failure of temporary oil transport pump which is installed thereafter is assumed. Because of this, all emergency D/Gs lose their function, and it causes the SBO.
2. Tsunami height between T.P. 17 m and T.P. 18 m
   Because, as shown by the fragility analysis result, the watertight doors of T/B and R/B are broken by tsunami of this height, inundation into the T/B and R/B occurs. Inundation into the T/B causes the flooding of support systems (e.g.,

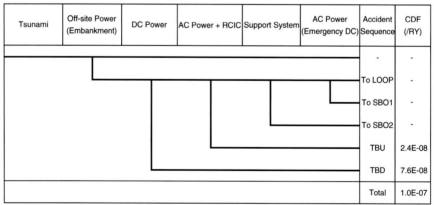

To LOOP: To child event tree for initiator event "Loss of Offsite Power", (the CDF of this sequence is 7.1E-11)
To SBO1: To child event tree for initiator event "Station Black Out with loss of emergency DC", (CDF: 2.8E-09)
To SBO2: To child event tree for initiator event "Station Black Out with loss of support system", (CDF:1.2E-09)
TBU: Transient with loss of all AC powers and ECCS(RCIC) injections
TBD: Transient with loss of all AC & DC powers

Fig. 10.7 Hierarchy event tree

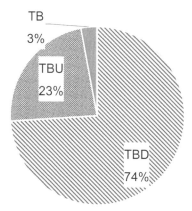

TBD: Transient with loss of all AC & DC powers

TBU: Transient with loss of all AC powers and ECCS(RCIC) injections

TB: Transient with loss of decay heat removal

Fig. 10.8 Contribution of each accident sequences for CDF in tsunami PRA (the state after the implementation of tsunami countermeasures)

RCW and RSW pumps) and the loss of its function, and then LUHS occurs. Also, inundation into the R/B causes the flooding of RCIC control panel and the loss of RCIC function. Then all of the water injection function failure is occurred.

3. Tsunami height exceeding T.P. 18 m
   Because, as shown by the fragility analysis result, the watertight door of C/B is broken by tsunami of this height, inundation into the C/B occurs, and it causes the loss of DC power (TBD).

Tsunami PRA result at the state after the implementation of countermeasures is shown in Fig. 10.8. Total CDF is calculated as 1.0E-7(/RY) in average value. As for accident sequence rate, TBD is dominant sequence accounting for 74 percentages in total CDF.

10.4   Effectiveness Evaluation About the Measure Taken in the Kashiwazaki-Kariwa Nuclear Power Plant

The validity of the measures against tsunami and power supply reflecting the lessons learned from the Fukushima Daiichi NPS accident will be evaluated by using the tsunami PRA. Here, the validity for the implemented safety measures is qualitatively discussed from the view of TQUV and TBD which are the important accident sequences determined prior to the implementation of additional safety countermeasures. Regarding TQUV, probability of LUHS and possibility of inoperability of RCIC by submersion will decrease due to installation of embankment, tidal wall, and watertight doors for important equipment rooms such as RCIC room and modification for maintenance hatch in T/B. Furthermore, even though all low-pressure water injection systems are lost by tsunami exceeding the embankment height, water injection can be done by fire engines located at high elevations. Therefore, in the state after the implementation of the tsunami countermeasures, it can be presumed that the occurrence probability of TQUV is reduced substantially. As for TBD, probability of LOOP and inoperable possibility of DC power by submersion will also decrease due to installation of embankment and watertight doors of important equipment rooms. In addition, the enhancement of DC power supplies is implemented for storage battery extension at higher floor in the reactor building, additional established storage battery, installation of the small generator, and maintenance of the DC power supply means. Accordingly, it is presumed that the possibility of loss of DC power decreases. Therefore, the present measures can be presumed as being appropriate against the important accident sequences extracted.

10.5   Conclusion

Tsunami PRA studies for Unit 7 of Kashiwazaki-Kariwa NPS was conducted, and the dominant accident scenarios that may result in core damage due to flooding were identified. The important accident sequences were evaluated as TQUV and TBD at the state before the implementation of countermeasures and CDF calculated as 2.1E-4 (/RY). This information supports qualitative assessment of the countermeasures that have been and will be implemented which indicates that these accident sequence probabilities will be decreased. Hence, the tsunami PRA was performed with the state after the implementation of tsunami countermeasures and CDF is calculated as 1.0E-7(/RY). By comparing these two CDFs, the effectiveness of the tsunami countermeasures which are implemented in the Kashiwazaki-Kariwa NPS is confirmed.

In this evaluation, there are some conservative assumptions, and total CDF is evaluated conservatively. However, for the purpose of safety enhancement, PRA should be implemented with more realistic method not to hide important sequences and equipment. With an emphasis on a sequence whose CDF is large, if the evaluation is impractical, there is a possibility that determination of safety measures will not be proper.

For example, countermeasures undetermined at that time of the PRA model design, such as waterproof treatment against internal flooding and drainage pumps, are not considered in the present evaluation. As a result, if outside watertight doors are broken due to wave pressure, the equipment in buildings is flooded and damaged according to the flooding depth. For the next step, more detail flooding propagation analyses in buildings with information of additional countermeasures are needed. The insights of inner flooding analysis will be also available. TEPCO will try to improve tsunami PRA continuously and enhance safety of the Kashiwazaki-Kariwa NPS using such results.

References

1. The standard of Tsunami Probabilistic Risk Assessment (PRA) for nuclear power plants: 2011, Atomic Energy Society of Japan, February 2012
2. Method of Probablistic Tsunami Hazard Analysis, Japan Society of Civil Engineers, March 2009

Chapter 11
Development of a New Mathematical Framework for Seismic Probabilistic Risk Assessment for Nuclear Power Plants – Plan and Current Status –

Hitoshi Muta, Ken Muramatsu, Osamu Furuya, Tomoaki Uchiyama, Akemi Nishida, and Tsuyoshi Takada

Abstract After the severe accident in Fukushima Daiichi Nuclear Power Station, safety improvement and enhancement have been installed. In midterm and long term, continuous efforts to improve and enhance safety are required, and technical basis and fundamentals are needed to achieve them.

Probabilistic Risk Assessment for seismic event (seismic PRA) is an effective measure to consider the countermeasures and improvement plans to secure the further safety of nuclear power plants regarding to seismic risk for the earthquake exceeding the design basis earthquake ground motion. However, the application of seismic PRA has not been utilized sufficiently so far. One of the reasons is that there is not enough agreement among stakeholders regarding to the evaluation methodology and consideration of uncertainty for decision-making.

This study proposes the mathematic framework to treat the uncertainty properly related to the evaluation of core damage frequency (CDF) induced by earthquake, the methodology to evaluate the fragility utilizing expert knowledge, the probabilistic model to cope with the aleatory uncertainty as well as the development of analysing code including these considerations for the improvement of the reliability of the methodology and enhancement of utilization of the products of seismic PRA.

This paper presents current status and some results from scoping calculations.

H. Muta (✉) • K. Muramatsu • O. Furuya
Department of Nuclear Safety Engineering, Tokyo City University, Setagaya-ku, Tokyo, Japan
e-mail: hmuta@tcu.ac.jp

T. Uchiyama
CSA of Japan, Minato-ku, Tokyo, Japan

A. Nishida
Center for Computational Science and e-Systems, Japan Atomic Energy Agency, Naka-gun, Ibaraki, Japan

T. Takada
Department of Architecture, The University of Tokyo, Bunkyo-ku, Tokyo, Japan

© The Author(s) 2016
K. Kamae (ed.), *Earthquakes, Tsunamis and Nuclear Risks*,
DOI 10.1007/978-4-431-55822-4_11

Keywords seismic PRA • Mathematical Framework • Uncertainty Analysis • High-Performance Computing • SECOM2-DQFM

11.1 Background

After the Fukushima Daiichi accident, safety enhancement of nuclear power plants in Japan is required by the new regulation. Moreover, continuous efforts to improve the reinforcement of risk management will be required in the middle or long term, and technical basis will be needed to support it. The importance of seismic PRA as a tool to identify potential accident scenarios caused by earthquakes, to estimate their likelihood and consequences and to support in assessing the effectiveness of measures to enhance safety against earthquakes has been widely and strongly recognized. However, seismic PRA has not been applied enough to achieve the aim above. One of the reasons is that there has been sufficient discussion and consensus building about the quantification and reduction of uncertainties in numerical results of seismic PRA and how to consider the uncertainty for decision-making.

In this study, a new mathematical framework of seismic PRA is proposed. Reviewing the current status of assessment procedures of accident sequence analysis in seismic PRA, this study will develop a new mathematical framework for estimating uncertainty in SPRA results in a more comprehensive way, taking into account uncertainties related to correlation effect of components failures which has been difficult to quantify so far. A computer code will be developed to materialize the proposed framework on the basis of the SECOM2-DQFM developed by JAEA to estimate the accident sequence occurrence probability and its uncertainty. The proposed mathematical framework is characterized by the following points:

- Representation of seismic hazard by a set of time histories of seismic motions using methods currently being developed by Nishida et al.
- Use of probabilistic response analysis by three-dimensional building model for determining responses of components to the seismic motions including the correlations among the component responses
- Use of Monte Carlo simulation for quantification of fault trees in accident sequence analysis
- Use of high-performance computing technology for realizing the use of above technologies in seismic PRA

Current status and some results from scoping calculations will be presented.

11.2 Current Framework and Challenges of Seismic PRA Methodology

In this chapter, firstly, general procedure and mathematical framework current method of seismic PRA should be reviewed; then issues of current uncertainty analysis framework will be extracted. Moreover, previous studies possibly to resolve the issues of mathematical framework.

11.2.1 Current Method of Seismic PRA

11.2.1.1 General Procedure of Seismic PRA

This study focuses the method of level 1 seismic PRA that evaluates the frequency of core damage accident. In general, the basic procedures of level 1 seismic PRA are shown in Fig. 11.1 and can be characterized as followings:

(a) Collecting the plant information and analyzing brief accident scenarios
To investigate the seismic source around the target site, characteristics of soil and structures, and safety system configuration, the brief accident scenarios induced by earthquakes are extracted.

(b) Seismic hazard analysis
Based on the information about faults around the target site and historical earthquake, occurrence frequencies of seismic ground motion exceeding a certain capacity such as maximum ground acceleration.

(c) Fragility analysis
To analyze the response and capacity of structures and components, the failure probabilities of structures and components can be expressed as fragilities, i.e. the function of capacity of seismic ground motion.

(d) Accident sequence analysis
To analyze seismic induced core damage accident sequences using event-tree (ET) and fault-tree (FT) techniques, core damage frequencies are evaluated based on these accident sequences, results of hazard analysis and fragility analysis.

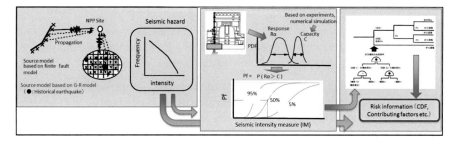

Fig. 11.1 Outline of seismic PRA methodology

11.2.1.2 Mathematical Framework of Current Method

In this study, focusing on the above items (c) and (d), mathematical model considering uncertainties of components and system failures will be studied. The mathematical framework for evaluating frequencies of accident sequences of seismic PRA is based on the concept by Kennedy et al. [1] and characterized as follows:

- The results of hazard analysis will be expressed as exceeding probabilities, that is, occurrence frequencies of seismic ground motions depending on the capacity on the target site. The levels of seismic ground motions are expressed as maximum accelerations of the surface.
- The wave used for response analysis is one of the time histories of waves such as design basis seismic ground motion. The impacts of variability of ground motion spectra are considered as variability of response factors explained later.
- The fragilities of components can be expressed as the probability that response exceeds capacity of the components, based on the assumption that probability distributions of response and capacity depending on the levels of seismic ground motion are the log-normal distribution, respectively.
- The median values of response depending on the seismic level are evaluated by linear extrapolation for the component response results associated with design basis seismic motion or interpolation of the results of calculations performed at several levels of ground motion.
- Standard deviations on the log scale for the response can be evaluated by expert opinion based on the results of the similar response analysis or comparison among observation points. Usually, response can be analyzed by the Sway-rocking model.
- Since responses are usually analyzed based on the design basis framework, response factors are introduced to consider impacts included in the assumption to secure conservatives of the design and to describe impacts of the uncertainty of model or data.
- Component capacities are expressed by median value and standard deviation; these parameters are set based on the results of structural analysis or verification test and, if necessary, expert opinion.
- Occurrence conditions of accident sequences are expressed as groups of minimal cut sets (MCS) equivalent to logical expression of accident conditions expressed by ET and FT. To calculate occurrence probabilities of these MCSs, the probability of certain accident sequence can be evaluated associated with the certain level of seismic ground motion.
- Core damage frequencies can be evaluated by the integration of the product of the probability of accident sequence associated with the certain ground motion level and seismic frequencies all over seismic ground motion levels.

Since the most important characteristics of the current framework is the extensive use of design information and the safety factors (response factors and capacity factors) that express the conservatism in the models used response and capacity evaluations in design, the current methodology is frequently called "the factors of safety method".

On the other hand, the method of Seismic Safety Margins Research Program (SSMRP) [2] is the other mathematical framework which adopts more detail model and input than that by Kennedy et al.; however, SSMRP method has not been fully used because of its complexity. Based on SSMRP method, new mathematical framework of this study will be established and presented in the next chapter.

11.2.2 Studies About Uncertainty Analysis Framework

11.2.2.1 Uncertainty Analysis Framework of Current Method

Current method was proposed to evaluate component failure probabilities by Kennedy et al. in 1980. The characteristics are as follows:

- Uncertainty of seismic hazard is expressed by the fractile curves that are composed of multiple curves corresponding to the percentage of the confidence level or aggregate curves corresponding to each set of alternative models and assumptions in calculating the hazard curve.
- Main causes of variability of model and data expressing response and capacity are categorized to "aleatory uncertainty" (or "uncertainty due to randomness") and "epistemic uncertainty" (or "uncertainty due to lack of knowledge"). The first one can't be reduced by the insights of experiments or theoretical studies because this type of variability is caused by inherent randomness of natural phenomena. The second one can be reduced by the insights of expansion of experimental data and enhancement of analysis models because this variability comes from lack of knowledge or simplification of analysis model.
- Usually, uncertainties in hazard analysis, fragility analysis, and in parameters of accident sequence models are propagated to the uncertainty in core damage frequency, while uncertainty of event tree and fault trees used in accident sequence analysis are considered by sensitivity studies.

11.2.2.2 Issues of Current Mathematical Framework

Seismic PRA is expected to provide useful insights and information for various decision-making. Important uses include the quantitative evaluation of the safety level of NPPs by comparing core damage frequency with quantitative safety goals and extraction of important accident sequences in a viewpoint of contribution to the total risk to enhance the safety features and accident countermeasures. So the followings are desirable and these needs are enhanced after Fukushima Daiichi accident:

- To reduce uncertainty in core damage frequency as far as possible.
- Plant damage states should be analyzed in detail. For example, how many systems failed simultaneously, how many structures such as buildings or piping failed or how they failed? What are the impacts of simultaneous occurrence of accidents in different units in a multiple unit site?

However, current seismic PRA method (the factor of safety method) has several difficulties that hinders improvement of its numerical accuracy; in other words resolution capability, for identifying important contributors, and many of them are tightly related to the simplification in the mathematical framework described above and are shown as follows:

(a) *Issues Mainly Related to the Hazard Analysis*

- The characteristic of seismic motion is expressed by only one parameter, i.e. peak ground acceleration. This means that dependency between the characteristic of the seismic source, i.e. distance and magnitude, and component failures is not modelled precisely enough.

(b) *Issues Mainly Related to the Fragility Analysis*

- Analytical models used in design calculations, for example, one-dimensional wave propagation model for the ground and Sway-rocking model for the building, sometimes may not be sufficient to provide detailed information to express the failure modes of structures and components precisely. The importance of the models to account for the three-dimensional response characteristics of the reactor building was pointed out after the Chuetsu-Oki earthquake in Japan.
- In such cases, building response models used in design have to be replaced by more detailed models such as three-dimensional finite element (3D-FEM) models. However, the use of such advanced models needs some additional efforts and information. It needs more detailed input seismic motion which should better be calculated from three-dimensional ground model using three-dimensional time history seismic motion data. This information is not provided from current framework of seismic hazard analysis. Furthermore, the parameters to express uncertainties in responses (the logarithmic standard deviation of response factors) have to be prepared. For Sway-rocking models, such parameters may be determined from existing studies. Since such preceding studies are not available for advanced models, some uncertainty calculations using the Monte Carlo or other statistical method will be necessary.

Although the required accuracy of response calculations in seismic PRA may not be as high as that required in deterministic safety assessments, it is desirable to have quantitative information on the impact of the differences in response calculations by different approaches.

(c) *Issues Mainly Related to the Accident Sequence Analysis*

- In case that response factor method is adopted in components response analysis, coefficients of correlation should be evaluated separately to consider the correlation of component response.
- In case that MCSs are used to evaluate core damage frequencies, since quantification considering simultaneous occurrence of multiple MCSs or dependency among multiple MCSs, error of calculation of core damage frequency tends to be increased.

- Since the range of correlation will be restricted, uncertainty of core damage frequency or contribution of each accident sequences tends to be increased.
- In case that initiating events are expressed by using hierarchical event tree, it is not obvious that the impacts of the simultaneous occurrences of multiple initiating events are considered sufficiently in the analysis. Moreover, since accident scenario analysis is very rough, resolution of the method could be reduced.

11.2.2.3 Previous Studies Possibly to Resolve the Issues of Mathematical Framework

Issues described above are possibly going to be resolved by the several previous studies. Individual insights and achievements to resolve the issues are the following:

(a) *Previous Studies Related to Hazard Analysis*

- Studies related to prediction of seismic motion regarding to modelling of seismic source using fault model, Green function method, semiempirical Green function method, and the combination of these methods can provide time historical wave considering seismic source characteristics.
- Nishida et al. proposed the method expressing seismic hazard by multiple time historical waves weighted by frequencies based on these above studies [3].

(b) *Previous Studies Related to Fragility Analysis*

- 3D response of structures and components evaluated by the techniques of structure response analysis such as finite element method (FEM) is gradually used to confirm the validity of seismic design.
- The enhancement of grid-computing method that makes high speed computing of structural response analysis possible using supercomputers makes large-scale FEM practical.
- Nishida et al. proposed the construction method of large-scale 3D plant model based on the structural analysis method, and it makes gradually possible the response analysis of major components of nuclear power plant using one linked model and the prediction of the failure point by detailed analysis of local stress of components.
- So many studies about probabilistic structural response analysis of components and structures have been done, for example, analysis of primary containment vessel by Takasaka et al. and failure probability analysis of piping system by Whitaker et al. Though preparation of time history wave associated with the level of seismic motion will be needed to link these insights to seismic PRA, however, those kinds of studies have not been done so far.
- For correlation of response of components, generic rule that describes how to evaluate the correlations among many components and to give the

coefficients of correlation considering the relationship among correlation of component, specific frequency of each component, and specific location in the building based on the probabilistic response analysis of structures was proposed in the SSMRP study. This is applied to the assessment of two nuclear power plants in NUREG-1150. Moreover, JNES, TSO of the former regulation body of Japan, studied to evaluate the correlation of response based on the soil-structure conditions using similar method and disclosed the results. These studies presented that it is possible to evaluate the correlation using probabilistic response analysis and implied that it could be possible to derive the rule to give the correlation coefficient from a series of detailed calculations in the simplified manner.

(c) *Previous Studies Related to Accident Sequence Analysis*

- Muramatsu et al. proposed the method that makes many samples of capacity and response by Monte Carlo simulation for quantification of FT in seismic PRA, named DQFM (direct quantification of fault tree using Monte Carlo simulation) [4, 5]. DQFM method is possible to quantify FT accurately better than MCS method and to consider the correlation of response among components in more general way. Moreover, SECOM2-DQFM that includes DQFM method is disclosed.
- DQFM method can calculate core damage frequency precisely even multiple initiating events occur simultaneously. So it could be useful to resolve the issue that accident sequence might be too much simplified by the hierarchical event-tree method if appropriate improvement is installed.

However, since huge efforts will be needed to make the mathematical treatment consistently from hazard analysis to accident sequence analysis to develop the new framework and method and to improve the whole mathematical method thoroughly in the application of insights and achievements of these above studies, current method has not been improved so far.

11.3 New Mathematical Framework for Seismic PRA Enhanced by High-Performance Computing

Based on the previous chapter, this study proposes brand new framework to resolve the issues above. This framework should be characterized by the following features:

1. *Seismic Hazard Analysis*

- Seismic hazard can be expressed by set of the groups including a set of seismic waves weighted by the occurrence frequencies based on the studies by Nishida et al.
- Uncertainty will be evaluated by expert opinion as necessary and expressed by the logic tree method.

2. *Fragility Analysis*

- Structures and soil are evaluated by 3D response analysis such as FEM or other method. It calculates a lot of cases associated with all of the set of seismic waves given to each level of hazard and uncertainty.
- Response and uncertainty of large-scale structures and components are analyzed coupling with building as a part of building response analysis.
- Floor response spectrum and its uncertainty of other than large-scale structures and components are analyzed using the results of building response analysis. Response and its uncertainty, i.e. median and log-scale standard deviation, are calculated using individual specific frequency and attenuation factor of each component.
- Current analysing method of capacity and its uncertainty of component are improved using study insights described in Sect. 11.3.2.

3. *Accident Sequence Analysis*

- Improving SECOM2-DQFM code that can use the results of 3D probabilistic response analysis based on the DQFM method, it is possible to analyze the conditioned core damage probabilities for each input time history seismic wave.
- Core damage frequencies are calculated to integrate the products of frequencies of occurrence of all of time history seismic waves and conditioned core damage probabilities, respectively.

These features are represented by the formula which is proposed by Sewell et al. [6] as follows:

$$\lambda[\Theta] \approx \sum_{All\ j} \Delta\{\lambda[x_j]\} \times \sum_{All\ k} P(TH_k|x_j) \times P[\Theta|TH_k, x_j] \qquad (11.1)$$

Here,

$\lambda[\Theta]$: Annual rate of the event
Θ: The event that some generalized "state of interest" is realized

$$\Delta\{\lambda[x_j]\}: \{\lambda[X \geq x_j]\} - \lambda[X \geq x_{j+1}]$$

X: A grand motion characterization
x_j: A specific value of interest at a site of interest
TH: Time history

As shown in Table 11.1, using 3D structure response analysis for fragility analysis, resolution, i.e. capability of scenario analysis, is enhanced significantly.

This framework requires large-scale calculations in the three fields such as composing a set of seismic waves of seismic hazard, large-scale probabilistic structure response analysis and quantification of system reliability model by

Table 11.1 Expected enhancement of resolution of seismic PRA by introducing 1D probabilistic response analysis

Related task of seismic PRA	Issues	Previous seismic PRA (AESJ standard and so on)	Improvements by probabilistic response analysis of whole plant using 3D FEM
Modelling of initiating events	Needed to consider multiple initiating events such as LOSP simultaneously. Current method considers the only single initiating event	Simplified by hierarchical event-tree method (Assuming capacity of structures such as Building > RPV > LOCA > Other events > LOSP. Superior events contain subordinate events. Conservative Evaluation)	Possible to consider multiple initiating events simultaneously by large-scale 3D plant model
Evaluation of failure probability of each component	Realistic analyses are required separately because capacities of components are evaluated conservatively in seismic design	For large-scale passive components, analyses in design stage or detail analyses are referred. Active mechanical and electrical components are analyzed by verification tests or vibration test results provided by the venders (Analysis methods are chosen depending on component types or availability of data.)	Basically, the same as the previous method, but detailed analyses can consider the diversity of spectrum characteristics of seismic motions and decrease the dependency on the decisions by analysts
Correlation among component failures	In analysis of simultaneous failure probability of multiplied systems, simultaneous failures of the same design components. i.e. consideration of correlations, are required	Quantitative evaluation of degree of correlation is difficult; simultaneous failures of the same design components are assumed conservatively	It is possible to rationalize the analysis of accident scenarios and the evaluation of CDF by introducing detailed response analysis method of whole structure that can evaluate precisely
Analysis of integrity of CV	It is important to analyze a location of CV failure for accident management	Detail analysis needed	The same as the left (It is possible to consider the diversity of spectrum characteristics of seismic motions)
Consideration of ageing effects	It is desirable to consider the impact of the ageing of component for countermeasures of ageing	Addressed as future work	It could be easier to evaluate the risk increase by reduction of capacity of components using 3D FEM

Monte Carlo method. It could be possible to realize considering the recent enhancement of supercomputing and expansion of inexpensive providing supercomputing.

To develop the analyzing system based on the concept of framework, the following two options are proposed:

11.3.1 Option A: Using High-Performance Computing Results Directly

Detail processes of this option are as follows:

(1) Seismic Hazard Analysis Including Uncertainty Analysis
   Seismic hazard is expressed by seismic motion that is described by a multiple set of seismic waves. However, to analyze uncertainty, each wave should include information of occurrence frequency, parameters of seismic source and propagation characteristics, uncertainty factor of those parameters such as occurrence probabilities, classification of aleatory and epistemic uncertainty.
(2) Soil-Structure Response Analysis Including Uncertainty Analysis
   Probabilistic response of soil structure is analyzed by 3D analyzing method such as FEM or Sway-rocking model that can treat 3D characteristics to some extent. In these analyses, factors of uncertainty and probabilistic distributions are determined by experts. Moreover, to calculate rationally, random variables treated in the analysis are focused on the dominant parameters. The results should contain the detailed location in the buildings, calculation input parameters such as occurrence probabilities and classification of aleatory and epistemic uncertainty.
(3) Accident Sequence Analysis Including Uncertainty Analysis
   Conditioned component failure probabilities and core damage probabilities are calculated using time history floor response obtained from soil-structure response analysis and component capacities for every time history data for seismic motions. In these analyses, uncertainties are analyzed as well using parameters for soil-structure response analysis.
(4) Uncertainty Analysis of CDF
   CDF and its uncertainty are calculated using frequencies of time history data for seismic motions and the results of the above item (3).

Figure 11.2 shows the process described above, and this process is named as "the direct method".

In some cases, this option requires more than 10,000 times of calculations of large-scale 3D structure response analysis, because it is needed to set probabilistic distributions for soil-structure parameters that can be focused on about 20 parameters, associated with 300 or more of time histories of seismic motions. It is possible to treat such size of calculations by simplification of 3D detailed model to some

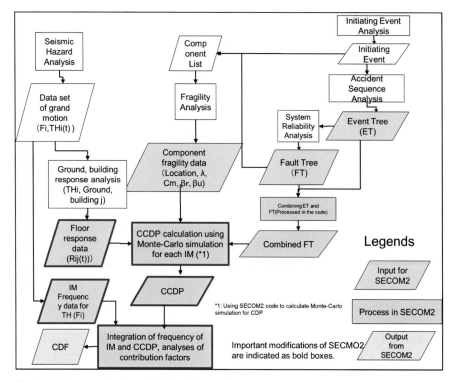

Fig. 11.2 Process of option A: Using high-performance computing results directly

extent and usage of supercomputers. However, since several sensitivity studies are required to analyze dominant factors, it is not practical.

11.3.2 Option B: Using Intermediate Parameters such as Capacity Factors Derived from Building Response Analysis

In seismic PRA process, so many trial and errors are needed to determine the analysis model, i.e. level of detail for system model and fragility. This means that the above "Option A" requires a lot of calculation times and is not practical. So combination of 3D analysis and response coefficient method is proposed as the intermediate method.

Preparing a number of calculations enough to simulate the probabilistic distributions of 3D analysis results at a certain degree of accuracy; then median, standard deviation and coefficient of correlation are determined to reproduce the results by response coefficient method using statistical analysis such as least-square method. Details of this method are as follows:

(1) Seismic hazard analysis including uncertainty analysis:
This is the same as "Option A".
(2) Soil-structure response analysis including uncertainty analysis:
This is the same as "Option A" as well.
(3) Derivation of seismic hazard curve indexed by maximum acceleration in free rock surface and response coefficient:
Calculation of the basis ground motion, seismic hazard curve associated with maximum acceleration in free rock surface, median and standard deviation of response coefficient, and coefficient of correlation among components are needed to apply the framework of seismic PRA using conventional response coefficient method. Response coefficient should be set to express the characteristic of probabilistic distribution very well.
(4) Accident sequence analysis including uncertainty analysis:
Using information determined in the above (3), component failure probabilities associated with the maximum acceleration in free rock surface and conditioned core damage probabilities are evaluated. Along with these analyses, uncertainty is calculated using parameters for soil-structure response analysis as well.
(5) Uncertainty Analysis of CDF:
CDF and its uncertainty are calculated based on the seismic hazard curve indexed by the maximum acceleration in free rock surface and conditioned core damage probabilities as well as uncertainties calculated in the above (4).

Figure 11.3a, b shows the optioned process related to the time history grand motions and floor responses described above, and these processes are named as "the intermediate method".

The most important point is that "response coefficient should be set to express the characteristic of probabilistic distribution very well by statistical analysis of the results of building 3D response analysis" described in item (3). This point is considered to be reasonable approximation if the three factors such as median, log-scale standard deviation and correlation are maintained properly in quantification process of CDF.

Based on this proposal, it could be expected that it is possible to model the more detailed 3D response characteristics of buildings and the more proper correlation that are the most important advantages of "Option A" by the practical calculation time.

However, this method is the intermediate and simplified method, and all of the advantages of 3D analysis could not be obtained. For example, the following issues need to be considered:

• Is it possible to introduce the index such as displacement, plastic deformation, other than acceleration in failure decision?
• Can it be suitably applied to the plastic region?
• Can the difference between the seismic source characteristics be reflected well in the calculation of the core damage frequency?

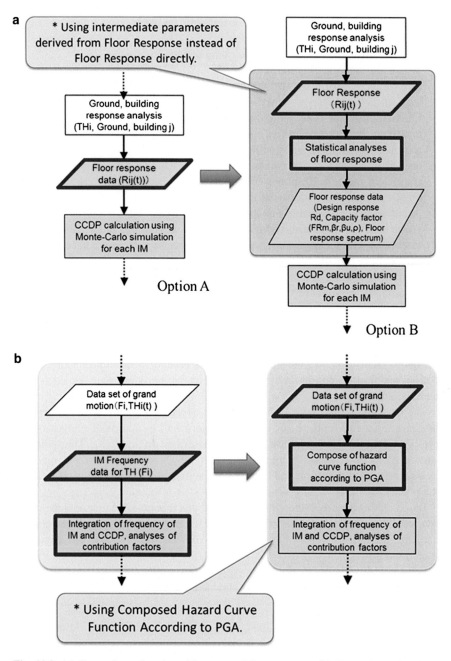

Fig. 11.3 (**a**) Comparison of options: Treatment of floor response (**b**) Comparison of options: Treatment of grand motion

To resolve these issues, it can be considered that a set of response coefficient associated with maximum acceleration level in seismic center or site will be introduced and so on.

11.4 Installation of Uncertainty Analysis Function Using Response Factor Method for SECOM2-DQFM

11.4.1 Improvement of SECOM2-DQFM CODE

The function of uncertainty analysis for core damage frequency (CDF), accident sequence frequency and intermediate event frequency is installed in SECOM2-DQFM. In detail, the followings are implemented:

- SECOM2-DQFM can be running on the large computing machine BX900 installed in JAEA.
- Improved to realize the large-scale grid computing.
- The uncertainty analysis function to calculate the uncertainty including aleatory uncertainty has been generally considered so far. The calculation time will be dramatically reduced by using hundreds of CPUs, even using proper set of random number for simulation.
- Improved to realize the uncertainty analysis of importance measurements such as FV importance.
- Improved to realize the uncertainty analysis, even considering the correlation among any events; that is one of the advantages of SECOM2-DQFM.

11.4.2 Analysis Results

Results of uncertainty analyses obtained by the improved and enhanced SECOM2-DQFM using the BWR5 model plant input are shown in Fig. 11.4.

The point estimate values and the mean value of uncertainty analyses are consistent, and this means that improvement of SECOM2-DQFM by this study is reasonable. From the uncertainty analysis results, 5 % lower value of 90 % confidence interval could not be obtained because these are too low to plot on the chart. The error factor of total CDF, which is derived from 95 % upper value of 90 % confidence interval divided by median value, is 11.0 and is smaller than those of each accident sequence, which is more than 10,000 in some cases. It is presumed that EF of the larger contributing accident sequences tends to be smaller than the smaller contributers relatively because of their small EFs.

Moreover, the EFs of lower frequency accident sequences are relatively larger, and the EFs of higher frequency accident sequences are relatively smaller. This is because that smaller contributing accident sequences include the components with

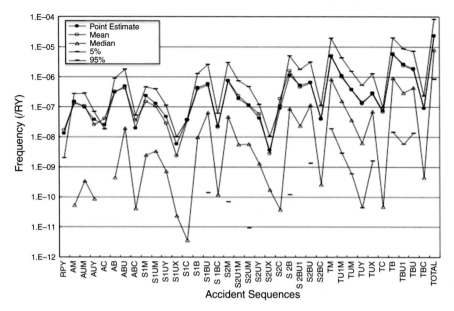

Fig. 11.4 Results uncertainty analyses for accident sequence frequencies

small fragility or redundancy. Especially, redundant components have complex relations of uncertainty, and these are cumulated in the calculation, and this is why the EFs of these accident sequences are so large.

11.5   Conclusions

A new framework is proposed to improve the resolution capability of seismic PRA. Improvement of computer code SECOM2 for quantification of FTs by Monte Carlo simulation is done. Based on these, capability of parallel processing was implemented to allow uncertainty analysis in a reasonable time for seismic PRA with the current model framework (response coefficient framework).

This study proposed the mathematic framework to treat the uncertainty properly related to the evaluation of core damage frequency induced by earthquake, the method to evaluate the fragility utilizing expert knowledge, the probabilistic model to cope with the aleatory uncertainty, as well as the development of analyzing code including these considerations for the improvement of the reliability of the method and enhancement of utilization of the products of seismic PRA.

Acknowledgements This study is performed as a part of a 3-year project "Reliability Enhancement of Seismic Risk Assessment of Nuclear Power Plants as Risk Management Fundamentals", which was started in 2012 and is funded by the Ministry of Education, Culture, Sports, Science and Technology (MEXT) of Japan.

References

1. Kennedy RP et al (1980) Probabilistic seismic safety study of an existing nuclear power plant. Nucl Eng Des 59:315–338
2. Smith PD et al (1981) Seismic safety Margins Research Program, NUREG/CR-2015
3. Nishida A et al (2013) Characteristics of simulated ground motions consistent with seismic hazard, SMiRT-22
4. Muramatsu K et al (2008) Effect of correlations of component failures and cross-connections of EDGs on seismically induced core damages of a multi-unit site. J Power Energy Syst JSME 2 (1):122–133
5. Kawaguchi K et al (2012) Efficiency of analytical methodologies in uncertainty analysis of seismic core damage frequency. J Power Energy Syst JSME 6(3):378–393
6. Robert T, Sewell et al (2009) Recent findings and developments in probabilistic seismic hazards analysis (PSHA) methodologies and applications, NEA/CSNI/R

Part IV
Nuclear Risk Governance in Society

Chapter 12
Deficits of Japanese Nuclear Risk Governance Remaining After the Fukushima Accident: Case of Contaminated Water Management

Kohta Juraku

Abstract It was found that many deficits of nuclear risk governance in Japan before and after the Fukushima accident. Not only were they created and embedded before the Fukushima disaster, but it has been remained or even worsened even after many accident reports were published and pointed out many problems and suggested ideas to remedy them.

In this paper, the author would analyze such remained problems found in the postaccident "on-site management" policy and measures, taking the case of contaminated water management at the Fukushima Daiichi Nuclear Power Plant. Firstly, the development of contaminated water management policy measures and institutional framework would be described in a chronological manner, which is one of the most typical and difficult tasks of "on-site management." Then, the cause of their failure trajectory would be analyzed by using a sociological concept "structural disaster" to understand the malfunctions which are continuously repeated not by identifiable particular factors but by inappropriate design of the socio-technical interface. This conceptual standpoint would suggest that the problems are not solvable by each of technical improvement, superficial institutional reform, nor prosecution and punishment of relevant individuals or organizations but by the redesign of that interface as a whole. Finally, based on this perspective, the author would discuss the ideas to remedy the deficits that might lead to further continuation of "structural disaster" in nuclear field.

Keywords Deficits of nuclear risk governance • Failure trajectory of postaccident on-site management • Contaminated water management at Fukushima Daiichi nuclear station • Structural disaster

K. Juraku (✉)
Department of Humanities and Social Sciences, Tokyo Denki University, 5 Senju Asahi-cho, Adachi-ku, 120-8551 Tokyo, Japan
e-mail: juraku@mail.dendai.ac.jp

© The Author(s) 2016
K. Kamae (ed.), *Earthquakes, Tsunamis and Nuclear Risks*,
DOI 10.1007/978-4-431-55822-4_12

12.1 Introduction: Failure Trajectory of Postaccident On-site Management

There were many deficits of nuclear risk governance in Japan before and after the Fukushima accident, as Taketoshi Taniguchi, the leading Japanese scholar in the field, illustrates by using the framework of "risk governance deficits," proposed by IRGC (International Risk Governance Council) [1, 2]. Not only were they created and embedded in the governance system before the Fukushima disaster, but it has been remained or even worsened after many accident reports were published and pointed out many problems which led the worst nuclear catastrophe in the history of Japanese nuclear utilization and suggested their recommendations to remedy them.

In this paper, the author would analyze such remained problems found in the postaccident "on-site management" policy and measures, taking the case of contaminated water management at the Fukushima Daiichi Nuclear Power Plant. "On-site management" includes many recovery works performed at the Fukushima Daiichi site—"stabilization" work in the language of the government and the Tokyo Electric Power Company (TEPCO)—such as setting up building covers to limit the further dispersion of radioactive substances; reinforcing buildings feared to have lost structural strength due to the effects of hydrogen explosions; containing contaminated water with various concentrations of radioactive substances, generated as a result of continuous water injection and cooling; and collecting and transporting leftover spent fuel.

It is one of the most difficult problems in the on-site management tasks mentioned above that the management of the highly radioactive contaminated water building up day by day. Due to various technical limitations, the temporary water injection and cooling system was built as not totally closed-cycle, and the damage caused by the accident allowed a huge amount of groundwater to flood the buildings. Inevitably, as the water gets contaminated with radioactive substances, highly radioactive contaminated water is continually produced. On top of this, since the path carrying the contaminated water to sea could not initially be identified or blocked, there were fears of marine pollution spreading.

Regarding this contaminated water treatment at the Fukushima site, a series of "follow-up" measures have been taken and caused the delay of underground water pump-out. Finally, the "ice wall" project to block underground water intrusion seems to be failed. Failures result in the increase of total amount of contaminated water and further social distrust about the feasibility and progress of Fukushima decommission project in Fukushima residents, entire Japanese society, and international community.

In the following sections, the author would describe the development of contaminated water management policy, measures, and institutional framework in chronological manner and then analyze the cause of their failure trajectory using a sociological concept "structural disaster" developed and suggested by Miwao Matsumoto, the pioneering sociologist of science who has shed light on the problems at the interfaces among science, technology and society, to understand the malfunctions which are continuously repeated not by identifiable particular factors but by inappropriate design of the socio-technical interface. This conceptual standpoint would suggest that the problems are not solvable by each of the technical improvement, superficial institutional reform, nor prosecution and punishment of relevant individuals or organizations but by the redesign of that interface as a whole [3, 4]. Finally, based on this perspective, the author discusses the ideas to remedy the deficits that might lead to further continuation of "structural disaster" in nuclear field.

12.2 Contaminated Water Management at Fukushima Daiichi Nuclear Power Plant

12.2.1 Failure to Build Consensus Through Explanations After the Fact and Follow-up Measures [1]: Delay in Addressing the "Groundwater Bypass" Problem

Thinking back now, more than 4 years since the accident, despite the common perception that the contaminated water problem only rose to prominence after the "acute phase" immediately following the accident, in reality the challenge of coping with the increasing volume of contaminated water was an agonizing problem in the locality from immediately following the accident. In fact, between April 4 and 11, 2011, lightly contaminated groundwater was released into the sea as a desperate, last-resort measure to secure space for highly radioactive contaminated water. This move was criticized by a number of neighboring countries. On May 11, 2011, only 2 months after the accident, a newspaper article described the seriousness of the contaminated water problem in a comprehensive manner [5].

As mentioned in that newspaper article, the factor that attracted attention as the biggest factor of the buildup of highly contaminated water was the problem of groundwater flooding [6]. The necessity for drastic measures to address this problem had already been recognized by the government and TEPCO in 2011 according to an official document of the first "Steering Meeting Under Government and TEPCO's Mid-to-Long-Term Countermeasure Meeting," but it was not until April 23, 2012, at the fifth "steering meeting," that the so-called "groundwater bypass" plan was officially presented. This was a detailed proposal to radically limit the buildup of contaminated water by pumping up groundwater before it could flood into nuclear reactor buildings to be contaminated by contact with radioactive substances. At the meeting, TEPCO presented a document titled *Use of Groundwater Bypass to Reduce Quantity of Groundwater Flooding into Buildings of Reactors Nos. 1 to 4*. TEPCO publicly announced anew its plan to pump up groundwater from before the flooding at a press conference on June 18, 2012 [7]. At the same time, TEPCO began providing explanations to fishing industry representatives, one of the major stakeholders. Since the problem of the release of the lightly contaminated water into the sea, mentioned above, the fishing industry representatives in Fukushima Prefecture became very sensitive about the contaminated water problem, so securing their agreement was vital to the success of TEPCO's plan. Even beyond the summer of 2012, TEPCO continued providing explanations to meetings of the association heads of the Fukushima Prefectural Federation of Fisheries Co-operative Associations ("Fukushima Fisheries Co-op").

Apparently, as a result of this process, in January 2013, the Fukushima Fisheries Co-op agreed to cooperate with TEPCO, reasoning that releasing groundwater was not the same thing as releasing contaminated water. Between that time, however, a leak of highly contaminated water occurred at the plant, and there were several suspected releases of water from the plant into the sea. This made the fishing industry representatives distrustful of TEPCO and led them to adopt a harder line in their negotiations. TEPCO proceeded to prepare facilities for their groundwater bypass, and they were ready to pump up groundwater and release it into the sea at any time, but a meeting of the association heads of the Fukushima Fisheries Co-op on May 13, 2013 decided to withhold its official agreement for a groundwater release [8]. If agreement had been obtained, TEPCO was set to start pumping up and releasing groundwater on the very next day, May 14, but it ultimately took another year or so before it could actually start releasing.

The reported reasons for withholding agreement to the plan was that a consensus could not be built among the co-op members, with members expressing views such as "only TEPCO is saying this, so we co-op members cannot trust them," and "we want TEPCO to clarify (explain to co-op members) that this is the national policy," according to the comments of co-op members cited in news paper articles [8, 9].

After this, a system was set up under which the responsible government body, the Agency for Natural Resources and Energy of the Ministry of Economy, Trade, and Industry (METI), and TEPCO would jointly provide explanations to stakeholders such as the fishing industry representatives. A number of explanatory meetings were subsequently held for fishing industry representatives and residents to gain the positive support to their groundwater bypass plan. Even though these efforts were made by them, however, multiple incidents of contaminated water leakage and newly discovered cases of water contamination were exposed after that, making it difficult to build a consensus.

Consequently, at an explanatory meeting organized by the national government and TEPCO for the Soma-Futaba Fisheries Cooperative Association ("Soma-Futaba Fisheries Co-op"), held on September 3, 2013, a chorus of criticisms about the release of groundwater was voiced. It was reported that the head of this association stated that "a decision on whether or not to agree to the bypass plan would be made no earlier than October, after gaging the reaction of distributors and consumers"[10].

On the same day as this explanatory meeting, the government's Nuclear Emergency Response Headquarters issued its "Basic Policy for the Contaminated Water Issue at the TEPCO's Fukushima Daiichi Nuclear Power Station." This policy provided for the setup of the "Inter-ministerial Council for Contaminated Water and Decommissioning Issues," the "Intergovernmental Liaison Office for

Contaminated Water and Decommissioning Issues," the "Intergovernmental Council for Fostering Mutual Understanding on the Contaminated Water Issue," and the "Fukushima Advisory Board Under the Council for the Decommissioning of the TEPCO Fukushima Daiichi Nuclear Power Station." In addition, the policy directed the national government to take, for the first time, direct financial measures toward contaminated water countermeasures (with provision for total funding of 47 billion yen). Then, the government and TEPCO continued working to provide explanations to stakeholders aimed at building a consensus, and on February 3, 2014, METI publicly disclosed "emission standards" for groundwater from the groundwater bypass, formulated jointly with TEPCO, stating that METI had explained the standards to the chairman of the Fukushima Fisheries Co-op.

Some time later, beginning in March 2014, efforts to reach a consensus intensified, but there were still further twists and turns in the process. On March 14, 2014, the government and TEPCO held an explanatory meeting for the Soma-Futaba Fisheries Co-op. Despite multiple expressions of opposition, the head of the co-op announced his approval, yet 4 days later, on March 18, the governing council of the Soma-Futaba Fisheries Co-op deferred a final decision on approval of the groundwater bypass plan. On the same day, the fishery co-op of Iwaki City decided to approve the plan at a meeting of its governing council. Finally, on March 24, 2014, the Soma-Futaba Fisheries Co-op officially issued its decision to approve the groundwater bypass plan, and on the following day, March 25, a meeting of the association heads of Fukushima Fisheries Co-op decided to approve the plan, with the submission of a request in writing to the government and TEPCO regarding the implementation of the plan.

Finally, on April 9, 2014, TEPCO began pumping up groundwater from wells, in accordance with the groundwater bypass plan, and on May 21, 2014, it released this groundwater (560 metric tons) into the sea for the first time.

So this groundwater bypass plan took two years to come to fruition, from the presentation of a detailed plan to the beginning of implementation. It is undeniable that the delay in executing the bypass plan to drastically control groundwater inundation greatly impacted the prospects for the overall success of the countermeasures to contain highly contaminated water. On August 2, 2013, the Nuclear Regulation Authority's (NRA) working group on contaminated water countermeasures pointed out that the groundwater level might rise suddenly as a sea-side impermeable wall was constructed and that even on completion, the outflow of contaminated water might not stop.

It should be viewed especially regrettable that by the spring of 2013, after having gone so far toward securing a final consensus from the fishing industry representatives, the most influential stakeholder, in fact consensus, could not be obtained and the process of building consensus process was carried forward anew.

12.2.2 Failure to Build Consensus Through Explanations After the Fact and Follow-up Measures [2]: Postponing "for the Time Being" the Response to the Problem of Releasing Lightly Contaminated Water into the Sea

One major factor that influenced the overall ins and outs of this story was the problem of releasing lightly contaminated water into the sea, which the author touched on earlier. The sudden buildup of contaminated water immediately following the accident rapidly caused a shortage of space to store contaminated water. As a result, in order to avoid highly contaminated water being inadvertently released into the sea, for a period of 1 week starting on April 4, 2011—approximately 3 weeks after the accident—lightly contaminated water was released into the sea to free up space to store more highly contaminated water. Given that this release was an emergency measure, the procedure for securing the approval of stakeholders was rather inadequate. As a result, there was criticism of the action from within and outside Japan, giving rise to a distrust of the government and TEPCO in relation to the handling of contaminated water.

Later too, the contaminated water storage capacity remained chronically insufficient, so from the same time as the abovementioned groundwater bypass plan, the idea of "releasing treated and purified contaminated water into the sea" was studied. However, in this case too, the views of stakeholders were not adequately reflected in the proposal. This adversely affected the effort to build a consensus on this later, delaying a response to the problem in terms of time.

When TEPCO publicly disclosed on December 8, 2011, that it was considering the release of treated and purified contaminated water into the sea, on the same day, Ikuhiro Hattori, the chairman of the National Federation of Fisheries Co-operative Associations ("National Fisheries Co-op"), visited TEPCO to express strong opposition to the proposal, calling it unacceptable. In the end, the idea of releasing treated and purified contaminated water into the sea was not included in the plan that TEPCO submitted to the Nuclear and Industrial Safety Agency (NISA) of METI on that same evening [11].

In a press conference on the same day, Nobutaka Tsutsui, Senior Vice-Minister of Agriculture, Forestry and Fisheries, also stated that the "release is unacceptable," indicating that TEPCO had publicly disclosed the plan without prior consultation with the major stakeholders.

As expected, in a plan submitted anew on December 15, 2011, TEPCO stated clearly that treated water "would not be released into the sea" by them for the time being [12]. Also on the same day, the Iwaki City Council in Fukushima Prefecture formally decided to request the repeal of the release plan.

Yet, according to some experts, the release of very lightly contaminated water within the limits of specified standards, with due consideration for risk management, is unavoidable. In a press conference on July 24, 2013, Shunichi Tanaka, the

Chairman of the Nuclear Regulation Authority, in reference to contaminated water within the limits of standards—not in reference to highly contaminated water that is treated and purified—stated that, "My frank opinion is that it's probably unavoidable to release a certain amount" [13]. Also, the review mission of the International Atomic Energy Agency (IAEA), which investigated the efforts to decommission the Fukushima Daiichi Nuclear Power Plant over 10 days, submitted a "summary report" to the Japanese government on December 4, 2013, with a recommendation that the controlled release of lightly contaminated water into the sea should be considered as an option [14].

While TEPCO put off the release of treated and purified contaminated water into the sea "for the time being," this effort cannot be easily excluded from a task list for appropriately managing the contaminated water problem. The fact that TEPCO initially tried to deal with the problem without securing a suitable consensus of stakeholders and that they took the easy option of deferring action "for the time being" in response to the opposition of stakeholders toward the issue may have considerably hindered the overall optimization of the contaminated water management. If TEPCO recognized that both the groundwater bypass plan and the plan to release treated and purified contaminated water into the sea were unavoidable and also that such countermeasures are more effective if taken promptly—and conversely, that they are unlikely to be effective and might even irreversibly aggravate the situation if not taken soon enough—it should have taken greater care in presenting its countermeasures in a form that ensures definite results, and even in the face of criticisms and doubts, it should have insisted on the necessity and effectiveness of the plans and the sufficiency of safety considerations, rather than simply "withdraw" or "defer" their plan. It is vital that TEPCO make decisions from a comprehensive perspective and with a clear commitment and that it presents its plans accordingly.

12.2.3 Incremental Development of a Governance System

TEPCO is not the only one grappling with these kinds of problems. The effort to construct a risk governance system for the Fukushima Daiichi Nuclear Power Plant, led by the government, could not be expected to be perfectly conceived from the start. By its very nature, it is an incremental development process.

The first platform set up by the government to comprehensively tackle measures aimed at decommissioning the plant, including contaminated water countermeasures, was the "Government and TEPCO's Mid-to-Long-Term Countermeasure Meeting," a coordinating body established on December 21, 2011. This body was set up based on an understanding that the situation would shift from a short-term recovery phase after the accident toward a medium to long-term decommissioning phase, in line with a declaration on December 16, 2011, by Prime Minister Yoshihiko Noda (then) about "recovery" after the accident. In response to a view of the Nuclear Emergency Response Headquarters that "in order to accelerate decommissioning, in addition to reinforcing research and development (R&D) systems focused on removal of fuel debris, it is important to construct a system to seamlessly manage on-site work and the progress of R&D," at the same meeting in February 2013 a decision was made to transform the coordinating body into the "'Council for the Decommissioning of TEPCO's Fukushima Daiichi Nuclear Power Station,' to include the heads of the main institutes engaged in R&D, in addition to the government and TEPCO" [15].

Later, in April 2013, the "Committee on Countermeasures for Contaminated Water Treatment" was set up under the "Council for the Decommissioning of TEPCO's Fukushima Daiichi Nuclear Power Station," to manage the planning and progress of government countermeasures to deal with the contaminated water problem. This committee put together a document, "Direction of Measures to be Taken (first round)," which was approved by the "Council for the Decommissioning of TEPCO's Fukushima Daiichi Nuclear Power Station" on June 27, 2013. This guideline was organized around three main pillars for action—introduction of a schedule for each nuclear reactor, enhancement of communication (through the "setup of the Fukushima Advisory Board Under the Council for the Decommissioning of TEPCO's Fukushima Daiichi Nuclear Power Station (tentative name)" for example), and "full-scale development of a system for gathering together international expertise." This third objective regarding "development of an international system" led to the establishment of the International Research Institute for Nuclear Decommissioning (IRID) on August 1, 2013. Furthermore, within the "Committee on Countermeasures," three "task forces" were set up between June and December 2013 to deal separately with each of these main challenges.

It is puzzling, however, that in December 2013 an "R&D Promotion Headquarters" was set up under the "Council for the Decommissioning of TEPCO's Fukushima Daiichi Nuclear Power Station," and in that case too four subordinate bodies were set up to undertake technical investigations—a "Working Team for Spent Fuel Pool Countermeasures," a "Working Team for Preparation of Fuel Debris Removal," a "Working Team for Radioactive Waste Processing and Disposal," and a "Joint Task Force for Remote Technologies."

On top of this, as already mentioned, in September 2013, the government's Nuclear Emergency Response Headquarters issued its "Basic Policy for the Contaminated Water Issue at the TEPCO's Fukushima Daiichi Nuclear Power Station," which called for the establishment of four subordinate bodies—the "Inter-ministerial Council for Contaminated Water and Decommissioning Issues," the "Intergovernmental Liaison Office for Contaminated Water and Decommissioning Issues," the "Intergovernmental Council for Fostering Mutual Understanding on the Contaminated Water Issue," and the "Fukushima Advisory Board Under the Council for the Decommissioning of the TEPCO Fukushima Daiichi Nuclear Power Station."

A "Decommissioning and contaminated water countermeasures team" was set up within the "Inter-ministerial Council for Contaminated Water and Decommissioning Issues" to investigate "studies of decommissioning and contaminated water countermeasures policy," "process management and risk clarification of decontamination and contaminated water countermeasures," "R&D needed for decommissioning and contaminated water countermeasures," and "collecting accurate information rapidly, providing it to residents, reporting it internationally, and addressing damage caused by rumors" [16].

The "Intergovernmental Council for Fostering Mutual Understanding on the Contaminated Water Issue," was set up at the same time, for the purpose of "enhancing information sharing in the locality" by TEPCO and the government regarding the contaminated water problem and the status of investigations into how to address the problem, "enhancing collaboration in the locality" between relevant bodies regarding contaminated water measures, and conducting "studies on how to proceed specifically with countermeasures, process management, and coordination between stakeholders."

The third new body, the "Fukushima Advisory Board Under the Council for the Decommissioning of the TEPCO Fukushima Daiichi Nuclear Power Station," was set up in February 2014 under the chairmanship of (then) State Minister of Economy, Trade and Industry Kazuyoshi Akaba (in charge of the abovementioned "Decommissioning and contaminated water countermeasures team"), with a membership including the deputy governor of Fukushima Prefecture; the heads of relevant municipalities in Fukushima Prefecture; representatives of local commerce, industry, agriculture, forestry, and fisheries interests; NPO representatives; and local community representatives.

In addition to all this, the NRA also set up its own "Specific Nuclear Facility Monitoring and Evaluation Committee" (set up in December 2012), along with a subordinate body, the "Working Group on Contaminated Water Countermeasures" (set up in August 2013), and also a "Team on Marine Monitoring" (set up in September 2013).

12.3 Discussion: Contaminated Water Management as a Case of "Structural Disaster"

Of course, the challenge of responding to this nuclear power plant accident was an extraordinary one. It would have been difficult to put in place any organizational system ahead of time, so this situation can be understood to be the outcome of the government proceeding flexibly to set up a system in accordance to the issues emerging along the way. Regrettably, however, there are too many deliberative bodies, and it is unclear how they all relate to each other. And even with this complicated system, it was not until 2013, approximately two years after the nuclear power plant accident, that the system was finally accelerated to be set up and operational. Considering, for example, that it was late 2011 when TEPCO presented and later withdrew its plan to release treated and purified contaminated water into the sea, as mentioned before, I can't help thinking that if at this point in time a system had been set up to enable TEPCO to work together with the government to pursue decision-making based on comprehensive investigation and coordination, and careful and open consensus building, the outcome could have been different.

This author's regret should not be considered as just a hindsight criticism. Rather, it must be understood as a result of the deficits of Japanese nuclear governance as Taniguchi demonstrates [1]. As mentioned earlier in this paper, his analysis adopts the "risk governance deficits" framework formulated by IRGC [2]. For example, the failures of contaminated water management described in this paper are counted as results of "Lack of adequate knowledge about values, risk perception, interests" deficit. He points out that another deficit, "Provision of biased, selective or incomplete information," is also found and the cause of it is something to do with the previous deficit. The problem is, thus, continuation or even reproduction of deficits after the accident. Why did the impact of worst accident not become an opportunity to stop it and change the Japanese nuclear governance better?

Matsumoto suggests a sociological concept that could shed light on the mechanism behind such persisting wrong trajectory: ""structural disaster" of the science-technology-society interface" [3, 4, 17]. This type of disaster is caused not by some failure of science, of technology or of society as separated manner. He argues, it should be considered as "the failure of the science-technology-society interface" [17]. There is no single technical failure, no obvious scientific misunderstanding, or no single person to be blamed. Rather, the interface among those heterogeneous elements of society as a whole suffers from serious problems. This understanding strongly suggests the possibility that "Efforts to pursue the perfect science cannot prevent the next problem. The perfect technology cannot, too. Society also cannot prevent it by ethical regulations" as Matsumoto points out. This perspective suggests us that the problems centering on Japanese nuclear policy and practices are not solvable by each of the technical improvement, superficial institutional reform, nor prosecution and punishment of relevant individuals or organizations but by redesign of that interface as a whole.

Of course, such a systematic view on technological failure has been developed, even before the Matsumoto's concept, for many years. There are many famous concepts to analyze it, such as "normal accident" [18], "organizational accident" [19], and so on. "Structural disaster" concept integrates such previous works and makes it clearer the conditions that cause the chain of accidents with similar characteristics.

According to Matsumoto, "structural disaster" consists of the following five elements [17]:

1. Following wrong precedents carries over problems and reproduces them.
2. Complexity of a system under consideration and the interdependence of its units aggravate problems.
3. Invisible norms of informal groups virtually hollow out formal norms.
4. Patching over problems at hand invites another patching over for temporary countermeasures.

5. Secrecy develops across different sectors and blurs the locus of agents responsible for the problems in question.

The author does not step into precise and point-to-point review of Fukushima contaminated water management case to determine if it meets the conditions above here, due to the limit of pages, but let him just point out some pertinent facts with those characteristics in the cases described in the previous section.

For example, the several causes of the delay of consensus-building and technical practice of the contaminated water treatment at Fukushima Daiichi site (both of groundwater bypass and lightly contaminated water release programs) can be considered as the cases of these elements. So-called "Kokusaku-Min-ei" (planned by the national Government, operated by private industry) scheme was not effective to gain public and stakeholders' trust for those measures, but TEPCO had acted as the front-end of those activities especially before the Governmental decision on September 2013. This fact can be interpreted as a result of elements 1 and 3. "Kokusaku-Min-ei" scheme was considered as the standard format of any nuclear activity.

This belief was strongly shared by many of the important stakeholders, such as the governments, TEPCO themselves, other member of industry, and even some of journalists and the general public. This seemed to be realized not by some formal consensus explicitly formed after the accident but by shared belief taken over from pre-Fukushima custom in nuclear industry in Japan. This point can also be interpreted as a sign of element 5, because the reason of "switch" of initiative from TEPCO to the government was not clearly discussed in public and explained well.

Also, too many relevant bodies and complicated network among them due to incremental development of the governance system for the contaminated water treatment can be seen as appearance of elements 2. Moreover, the "for the time being" strategy is a strong sign of element 4, of course.

In this way, the twists and turns story described in this paper shows many signs of those five conditions. It is obvious that the deficits identified by Taniguchi seem to be strongly related to the mechanism of "structural disaster."

12.4   Concluding Remarks: To Remedy Structural Deficits of Japanese Nuclear Governance

In light of the discussion above, it can be said that sociological analysis of mechanism behind the series of problems of post-Fukushima accident on-site management should be important and prospective to think about the remedy for its failure trajectory, although the author could not demonstrate the result of detailed analysis in this paper.

Of course, it is essential to promote technical R&D to deal with contaminated water better. It should be useful to solve many difficult problems at the damaged plant site. It is also critical to establish appropriate institutional and legal framework to support those activities.

However, even such effort might become a part of next "structural disaster" if we don't have deliberate and proper understanding on the mechanism that creates the chain of accidents, incidents, and scandals. "Quick fixes" for superficial layer of problems often make the problems more complicated, unsolvable, and serious. Sociological perspectives should be able to make contributions to avoid it and to enrich our wisdom to tackle the deficits. As an idea for this, Matsumoto suggests his solution for "structural disaster" that includes the introduction and establishment of plural channels among science-technology-society by "position-indicated style" interpreters and research funding scheme to enable open, transparent, and responsibility traceable policy (it is the opposite to the faulty one that create "structural disasters").

We can collaborate to stop the chain of "structural disasters" by considering such proactive suggestion from sociologist as well as other social scientist in various fields. The problem of structural deficits of Japanese nuclear governance can and should become the good and pioneering example of interdisciplinary collaboration between engineering and sociology (and other social sciences). It should be enhanced and promoted more immediately.

Acknowledgements Part of this paper is based on the author's book chapter written in Japanese [20] and supported by the JSPS (Japan Society for Promotion of Science) academic funding program "Higashi-Nihon Dai-shinsai Gakujutsu Chousa" (Academic Survey Program for Great East Japan Disaster).

References

1. Taniguchi T (2014) Lessons learned from deficits analysis of nuclear risk governance. International symposium on earthquake, tsunami and nuclear risks after the accident of TEPCO's Fukushima Daiichi Nuclear Power Stations, Kyoto University, 30 October 2014, Kyoto, Japan
2. IRGC (International Risk Governance Council) (2010) Risk governance deficits: an analysis and illustration of the most common deficits in risk governance. International Risk Governance Council, Geneva
3. Matsumoto M (2002=2012) Chi no Shippai to Shakai: Kagaku-Gijutsu wa Naze Shakai ni Totte Mondai-ka [The structural failure of the science-technology-society interface]. Iwanami Publishing Co., Tokyo (in Japanese)
4. Matsumoto M (2012) Kozosai: Kagaku-Gijustu-Shakai ni Hisomu-Kiki [The structural disaster: crisis hidden in techno-scientific society]. Iwanami Publishing Co., Tokyo (in Japanese)
5. Asahi Shimbun (2011) Osen-sui, Kuno 9-man ton: Kensho, Fukushima Daiichi Genpatsu no Chusui to Taio [Contaminated water, agonizing 90 thousand tons: review of water injection and the other counter measures at Fukushima Daiichi Nuclear Power Station], 11 May 2011. (in Japanese)

6. Nuclear Emergency Response Headquarters (2013a) Basic policy for the contaminated water issue at the TEPCO's Fukushima Daiichi Nuclear Power Station, decided on 3 September 2013. http://www.meti.go.jp/english/earthquake/nuclear/decommissioning/pdf/20130904_01a.pdf

7. Tokyo Electric Power Company (TEPCO) (2012) Genshi-ro Tate-ya-to e-no Chika-sui Ryunyu ni-taisuru Bapponteki-Taisaku no Kento Jokyo ni-tsuite [The examination progress of fundamental countermeasures against water flooding into reactor buildings and other facilities], released on 18 June 2012. (in Japanese)

8. Asahi Shimbun (2013a) Chika-sui Housyutsu, Ryoshi no Fushin: Touden-no Keikaku, Fukushima-ken Gyoren wa Ryousyo-sezu [Ground water discharge, distrust of fishermans: TEPCO's plan was not approved by Fukushima Prefectural Fishery Co-op], 14 May 2013. (in Japanese)

9. Fukushma Minpo News (2013) Ken-Gyoren Kaiyo-housyutsu-ni Fudo-i: Daiichi Genpatsu Chika-sui Anzen-e-no Kenen Fussyoku Sarezu [Disagreement of Prefectural Fishery Co-op: ground water at Fukushima Daiichi NPP, Concern about Safety was not Settled], 14 May 2013. (in Japanese)

10. Asahi Shimbun (2013b) Bai-pasu Handan Miokuri: Sou-Sou Gyokyo, Shiken-Sogyo Saikai wo yusen, Osen-sui Taisaku Setsumei-kai [Decision on the By-pass Plan Postponed by Sou-Sou Fishery Co-op at the briefing session of Contaminated Water Management: trial operation of fishing comes first] 4 September 2013. (in Japanese)

11. Tokyo Electric Power Company (TEPCO) (2011a) The report of operation and management plan for our facilities based on "Policy on the mid term security" for the Units 1 to 4 of Fukushima Daiichi Nuclear Power Station(2), 12 December 2011. (in Japanese)

12. Tokyo Electric Power Company (2011b) The report of operation and management plan for our facilities based on "Policy on the mid term security" for the Units 1 to 4 of Fukushima Daiichi Nuclear Power Station(3), 15 December 2011. (in Japanese)

13. Nuclear Regulatory Agency (NRA) (2013) Press Conference Note, 25 July 2013 (in Japanese)

14. International Atomic Energy Agency (IAEA) (2013) Preliminary summary report IAEA international peer review mission on mid-and-long-term roadmap towards the decommissioning Of TEPCO's Fukushima Daiichi Nuclear Power Station Units 1–4 (Second Mission)

15. Nuclear Emergency Response Headquarters (2013b) Tokyo Denryoku Fukushima Daiichi Genshi-ryoku Hatsuden-sho no Hairo-taisei no Kyouka-ni-tsuite [Strengthening the system for decommission of TEPCO's Fukushima Daiichi Nuclear Power Station], decided on 8 February 2013. (in Japanese)

16. Nuclear Emergency Response Headquarters (2013c) Strengthening the system for addressing contaminated water and decommissioning issues at the TEPCO's Fukushima Daiichi Nuclear Power Station, decided on 10 September 2013. http://www.meti.go.jp/english/earthquake/nuclear/decommissioning/pdf/20130910_01a.pdf

17. Matsumoto M (2013) "Structural disaster" long before Fukushima: a hidden accident. Dev Soc 42(2):165–190

18. Perrow C (1984=1999) Normal accidents: living with high-risk technologies. Princeton University Press, Princeton

19. Reason J (1997) Managing the risks of organizational accidents. Ashgate Publishing Ltd, Aldershot

20. Juraku K (2015) Section 4: Post-accident on-site management, a section of Chapter 6: Post-accident nuclear power technology governance. In: Shiroyama H (ed) Supervised by M. Muramatsu and K. Tsunekawa, Dai-shinsai-kara Manabu Shakai-kagaku (Vol.3) Fukushima Gen-patsu Jiko-to Fukugou Risuku-Gabanansu (Social science learnt from the great earthquake (Vol.3) Fukushima Nuclear Accident and Complex Risk Governance). Toyo Keizai Inc., Tokyo (in Japanese)

Chapter 13
A Community-Based Risk Communication Approach on Low-Dose Radiation Effect

Naoki Yamano

Abstract A community-based risk communication approach for risk and risk-related factors regarding low-dose radiation has been started in 2013. In this approach, three groups that consist of local citizens, health nurse, midwife and nutritionist, and media reporters have been coordinated, and they discuss and examine a guidebook of health effects on low-dose radiation prepared by experts. Then they will revise the contents and expressions of the guidebook under expert's support by themselves. An improved guidebook implementing stakeholders' input will be expected through this process. In parallel to the community-based risk communication, an opinion survey has been conducted for the inhabitants of Tsuruga City in the Fukui prefecture to obtain people's cognition about ionizing radiation and risk on health effects. The inhabitants of about 43 % show strong anxiety for radiation. They also show strong requirement for the risk criteria that should be specified by government/specialists. The current status and progress of the community-based risk communication approach are discussed, and a necessity of risk education regarding trans-science problem is presented.

Keywords Risk communication • Low-dose radiation • Fukushima nuclear accident • Public engagement

13.1 Introduction

Even now, after three and a half years or more from the Fukushima Daiichi nuclear accident, the health effects of low-dose ionizing radiation have become a national anxiety. Many activities of risk communication performed by the government are likely not successful though most inhabitants received estimated dose less than a few mSv.

Before the Fukushima accident, nuclear risk communication in Japan has been developed for public acceptance and improvement of nuclear power promotion understanding under the prerequisite that the safety is ensured. The risk

N. Yamano (✉)
Research Institute of Nuclear Engineering, University of Fukui, 1-2-4 Kanawacho, Tsuruga, Fukui 914-0055, Japan
e-mail: yamano_n@u-fukui.ac.jp

© The Author(s) 2016
K. Kamae (ed.), *Earthquakes, Tsunamis and Nuclear Risks*,
DOI 10.1007/978-4-431-55822-4_13

communication is expected as an effective method to regain public trust after the Fukushima accident. There is a good review for the historical background of risk communication in Japan [1]. However, risk communication for the health effects of low dose of ionizing radiation is not easy to perform because the low-dose effect has uncertainty including its sociological and psychological nature. Many specialists including radiation scientists, biologists, and health physicists have tried to explain the low-dose radiation effect to public, but it cannot be said that the public understanding through dialogue is effective because the specialists are not communication experts. The difficulties of risk communication for low-dose radiation have been reviewed by Kanda [2].

The author has started a new risk communication approach concerning health effects of low-dose radiation in 2013. This approach is intended to establish a community-based risk communication regarding low-dose radiation. In parallel to the community-based risk communication, an opinion survey has been done at September 2013 for inhabitants of Tsuruga City in the Fukui prefecture where 14 nuclear reactors are located in its vicinity. In Chap. 2, a summary of the opinion survey and the result is described. Chapter 3 describes the community-based risk communication approach and its progress. Chapter 4 will show insights and discussion through the community-based risk communication approach.

13.2   Opinion Survey for Tsuruga Inhabitants

The opinion survey was conducted at mid-September 2013 for 300 adult inhabitants of Tsuruga City in the Fukui prefecture. The aim of this survey is to obtain people's cognition about ionizing radiation and risk on health effects.

The survey method was a visit questionnaire placement method to obtain a sample number of 300 by assigning 15 people each in residential areas of 20. Questions consisted of (1) the awareness about radiation and radioactivity, (2) the awareness about "risk" and food safety, and (3) the awareness about health effects of low-dose radiation exposure. In this study, the authors avoided direct questions concerning awareness of nuclear power. This indicates that it is not a survey questioning the pros and cons of nuclear power. Once people recognized it as a questionnaire relating to approval or disapproval of nuclear power, there is a possibility that influence of bias occurs in answer.

The detailed analysis of the opinion survey is described elsewhere [3].

In terms of risk cognition, Tsuruga inhabitants have the following thoughts about risk:

- There is a correlation between experience of risk education and knowing how to judge risk.
- There is a correlation between knowing the meaning of risk and judging own risk.
- There is a tendency that knowing how to judge risk leads a sense of security.

- People do not recognize a risk cognition which has a trade-offs relation between hazard and benefit.
- There is no correlation between experience of risk education and judging own risk.
- The requirement of Tsuruga inhabitants for the risk acceptance criteria of low-dose radiation that should be specified by government/specialists does not depend on the degree of their knowledge about the risk information of radiation.

The results are obtained from the detailed analysis of the opinion survey, so we indicate the reference [3]. The tendency of these correlations about risk knowledge and recognition shows some people have experience of risk education in specific field, but the knowledge is not for radiation. It is clearly shown that inhabitants have no experience of risk education on low-dose radiation.

13.3   Community-Based Risk Communication Approach

In terms of low-dose radiation, most of the general public have few knowledge on what is a probabilistic (stochastic) effect on radiation exposure and the risk. The risk concept is not uniquely defined yet. Some scientists and engineers have recognized that risk is "probability × consequences." However, according to the ISO guidelines on risk management ISO31000:2009 [4], the definition is "effect of uncertainty on objectives." The International Risk Governance Council (IRGC) also defines risk as "an uncertain consequence of an event or an activity with respect to something that humans value" [5]. Such consequences can be positive or negative, depending on the values that people associate with them.

There is some confusion about risk concept among specialists in different area. Also, a consensus of the stochastic health effects of low-dose radiation has not been established among radiation scientists and biologists.

After the Fukushima Daiichi accident, many Japanese radiation scientists tried to explain the low-dose radiation effect to the general public. A lot of risk information on health effects of radiation has been explained to the general public using persuasive message based on epidemiologic study and the LNT (linear no-threshold) model. The general public has anxiety for radiation, so the question to scientists is simple such as "Am I safe? What is health impact to children and offspring?" However, the answer is not simple because of the uncertainty of scientific evidence from the epidemiologic study and the LNT hypothesis. It shows the health effects of low-dose radiation mean a trans-science problem.

General public is not familiar with the "probability." They also have heard different opinions for the "probability effect" of the low-dose radiation from some radiation scientists or commentators. So, the majority of the general public feels the scientists and commentators untrustworthy. In this situation, it means that the risk communication for low-dose radiation is not easy to perform because of lack of credibility of scientists.

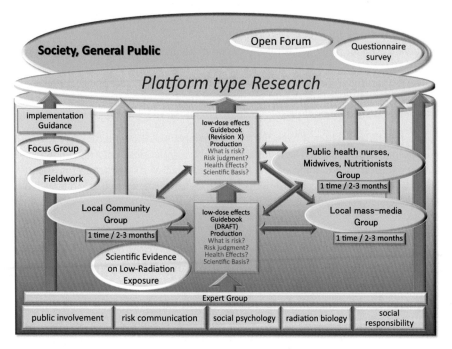

Fig. 13.1 A concept of the community-based risk communication approach

In general, a risk communication method which is designed for promoting stakeholders' willingness concerning "the right to know" and "the right to self-determination" is effective. There is diversity in the values of the general public. There are stakeholders who think that government/specialists should specify the risk criteria, but some stakeholders think strongly and want to judge it on their own based on the right to self-determination.

Based on this insight, a new risk communication approach concerning health effects of low-dose radiation has started in FY2013. A conceptual diagram of this approach is shown in Fig. 13.1. This approach is intended to establish public engagement of risk communication for risk and risk-related factors regarding low-dose radiation to promote the right to self-determination.

Firstly, a draft of guidebook of health effects on low-dose radiation was prepared in cooperation with experts of radiation biology, social psychology, risk communication, public involvement, and social responsibility. The draft consists of 46 pages including six chapters and references. The contents are Introduction (1), Concept of risk (2), Deterministic effects on radiation (3), What is the low dose? (4), Low-dose effects on radiation (5), and Conclusion (6).

Secondly, three groups have been coordinated in cooperation with Tsuruga inhabitants as follows:

- Local community inhabitants (12) who are members of female study group on nuclear power

Fig. 13.2 Participants of workshop, (**a**) local community inhabitants; (**b**) health nurse, midwife, and nutritionist; (**c**) media reporters

- Health nurses, midwives, and nutritionists (12) who belong to the Tsuruga City health care center
- Media reporters (5) who belong to the Tsuruga press club

where the number in parenthesis indicates the number of participants in each group.

Participants of the three groups discuss and examine the draft of guidebook in workshop as shown in Fig. 13.2. Based on the dialogue and consultation, they will revise the contents and expressions of the guidebook under specialist's support by themselves. An improved guidebook implementing stakeholders' input will be expected through this process. The workshop has been held eight times in FY2013. In FY2014, creating the guidebook for beginner is in progress.

Besides that, we held public forums in the Tokyo metropolitan area and Tsuruga for the related researchers and general public. There were discussions on the opinion of the participants about the risk communication approach.

13.4   Discussion

Through the workshop using the draft of guidebook of health effects on low-dose radiation, the following opinions were obtained from participants:

- It is difficult to understand the risk concept, e.g., definition, probability, and uncertainty of risk.
- It is difficult to understand jargons such as DNA damage, repair mechanism, LNT model, EAR, ERR, Sv, Gy, or Bq.
- Epidemiologic study on health effects of low-dose radiation like CT scan is difficult to understand, but psychological impact after Chernobyl accident is well understood.
- There is too much information to understand. Avoid detailed contents, and fewer pages are favorable.
- It should change the order of the chapter because the risk concept is difficult.
- It is favorable to create a beginner's guidebook with fewer pages.

Fig. 13.3 Trade-offs relation between hazard and benefit on radiation risk

It seems that the general public does not easily understand the risk concept including definition and the risk cognition. Risk can be recognized as positive or negative depending on the values that people associate with them. If ionizing radiation is hazardous and has no benefit to people, they will not accept it whether it is low dose or not.

So, a relationship of trade-offs between hazard and benefit is important to understand radiation risk. Authors describe risk can be recognized as positive and negative, and the relationship between hazard and benefit as negative versus positive. We are trying to explain the risk trade-offs using a balance scale between (negative) hazard and (positive) benefit as shown in Fig.13.3.

Potential risk of low-dose ionizing radiation is "hazard" as cause of cancer, but it also has "benefit" like early detection of disease. The meaning of radiation risk and how to judge the risk on own is important to know. Most of the general public has no experience on learning about risk at education institutions. Some Tsuruga inhabitants have learned about risk at education institutions, but it seems that the risk education was for typical application, not for low-dose ionizing radiation. Even though they have knowledge on how to make typical risk judgment to other applications, they believe that the knowledge cannot be applicable to their health effects of ionizing radiation.

Through workshop, participants discussed with each other issues concerning the importance of understanding health effects of radiation and how to judge the radiation risk. They have understood that they should judge the risk of low-dose radiation on own whether the risk criteria specified by government/specialists are adequate or not. They also understood how to avoid or reduce the risk of artificial ionizing radiation.

13.5 Conclusion

A community-based risk communication approach concerning health effects of low-dose radiation has been described. The method is designed to promote stake-holders' willingness concerning not only "the right to know" but also "the right to self-determination."

The author conducted an opinion survey for Tsuruga inhabitants in order to obtain people's cognition about ionizing radiation and risk on health effects. The inhabitants of about 43 % people show strong anxiety for radiation and have strong requirement for the risk criteria that should be specified by the government/specialists. It is believed that they are not satisfied with the current criteria of low-dose radiation. From the opinion survey, most of the public have no experience learning about the risk at education institution. However, it is found a tendency that knowing how to judge risk leads a sense of security. So, risk education for low-dose radiation seems to be able to reduce unnecessary anxiety.

The current status and progress of the community-based risk communication approach have been discussed. Through the community-based risk communication approach, participants have understood that they should judge the risk of low-dose radiation by their own, whether the risk criteria specified by government/specialists are adequate or not. Participants also understood how to avoid or reduce risk from artificial radiation. An improved guidebook implementing stakeholders' input will be expected through this process. The present method will be effective to public understanding about risk of low-dose radiation.

Acknowledgments This work was supported by JSPS KAKENHI Grant Number 25420902.

References

1. Kinoshita T (2014) Short history of risk communication in Japan. J Dis Res 9:592–597
2. Kanda R (2014) Risk communication in the field of radiation. J Dis Res 9:608–618
3. Shinoda Y, Yamano N (2015) Survey of Tsuruga inhabitants concerning radiation and its risks. Trans Atomic Energy Soc Jpn 14(2):95–112, doi:10.3327/taesj.J14.018. [in Japanese]
4. ISO 31000:2009, Risk management – principles and guidelines. 2009. International Organization for Standardization
5. Renn O (2005) White paper on risk governance, towards an integrative approach. International Risk Governance Council. Available at http://www.irgc.org/